普通高等教育土木与交通类"十三五"规划教材

岩土工程概论（中英双语）

主　编　刘爱华
副主编　唐丽燕　李　青　赖佑贤

·北京·

内 容 提 要

本教材既考虑岩土工程课程专业知识上的系统性,也兼顾科技英语学习的基本特点与要求,通过优化教材内容,实现专业知识的学习和科技英语能力的培养提高,达到双赢目标。内容涵盖岩土工程的基本分类方法与典型岩土工程,岩土材料的力学特性和测试方法,岩土工程相关计算理论、设计方法、施工技术和装备,岩土工程常见工程灾害与防治,数值化技术与岩土工程的关系以及学科领域的研究热点和前沿等内容。章与章之间既相互联系,又可自成一体,方便满足不同层次读者的需求。

本教材可作为高校土木、水利、建筑、交通及其他岩土工程类专业学生的双语教材,也可作为相关学科领域研究生以及工程技术与科研人员的学习参考资料。

图书在版编目(CIP)数据

岩土工程概论:汉、英 / 刘爱华主编. -- 北京:中国水利水电出版社,2016.11
普通高等教育土木与交通类"十三五"规划教材
ISBN 978-7-5170-4915-9

Ⅰ.①岩… Ⅱ.①刘… Ⅲ.①岩土工程-高等学校-教材-汉、英 Ⅳ.①TU4

中国版本图书馆CIP数据核字(2016)第280665号

书 名	普通高等教育土木与交通类"十三五"规划教材 **岩土工程概论(中英双语)** YANTU GONGCHENG GAILUN (ZHONGYING SHUANGYU)
作 者	主编 刘爱华 副主编 唐丽燕 李 青 赖佑贤
出版发行	中国水利水电出版社 (北京市海淀区玉渊潭南路1号D座 100038) 网址:www.waterpub.com.cn E-mail:sales@waterpub.com.cn 电话:(010)68367658(营销中心)
经 售	北京科水图书销售中心(零售) 电话:(010)88383994、63202643、68545874 全国各地新华书店和相关出版物销售网点
排 版	中国水利水电出版社微机排版中心
印 刷	北京瑞斯通印务发展有限公司
规 格	184mm×260mm 16开本 15印张 397千字
版 次	2016年11月第1版 2016年11月第1次印刷
印 数	0001—3000册
定 价	**38.00元**

凡购买我社图书,如有缺页、倒页、脱页的,本社营销中心负责调换
版权所有·侵权必究

主 编 简 介

刘爱华　男，1963 年 3 月生，湖南邵东人，博士、教授、博士研究生导师。国际岩石力学学会会员，中国岩石力学与工程学会工程实例委员会委员，企业安全生产标准化建设指导与评审专家。长期从事岩石力学计算方法与理论、岩土工程加固机理与灾害防治以及安全管理工程等领域的教学与科研工作。

1983 年毕业于中南矿冶学院（中南大学），1988—1995 年公派留学法国，获国立巴黎高等矿业学校硕士和博士学位，并在巴黎第十二大学完成博士后工作。留学法国期间，主要研究方向为岩石力学与工程地质理论及其应用。1995—1997 年任职中南工业大学（中南大学），1996 年晋升为教授。1997—2004 年在美国工作，曾就职于 CPI 和 IBM 等公司。2004—2013 年任中南大学教授、博士研究生导师，2013 年 12 月调入华南农业大学水利与土木工程学院。

曾获教育部 1995 年度"优秀青年教师资助计划"、国家教委优秀留学人员基金、IET 等多种奖励。主持或承担过多项基金研究和横向科研项目，曾担任"973"项目子课题负责人 1 次以及"973"项目研究骨干 1 次；科研成果获得过省部级科技进步特等奖、一等奖、二等奖等奖励；教学研究成果获得过校级教学成果一等奖及省级教学成果二等奖。已出版著作 4 部，发表科研论文 80 多篇，获国家发明专利授权 1 项。

E-mail: alexliu@163.com

Editor in chief

Liu Aihua, male, born in March 1963, Shaodong, Hunan, is a Ph. D, Professor, Ph. D tutor. Professor Liu is a member of the International Association of Rock Mechanics, committee member of the Real Engineering Cases Association of Rock Mechanics and Engineering in China, and expert both in construction and evaluation of enterprise safety production standardization. His major research fields include calculation methods and theories of rock mechanics, reinforcement mechanisms of rock and soil engineering, and safety management engineering.

Mr. Liu graduated from Central South Institute of Mining and Metallurgy (Central South University) in 1983 and obtained a Bachelor's Degree in Mining Engineering. Mr. Liu had a long experience of overseas studies. He studied in Ecole Nationale Superieure des Mines de Paris, France, from 1988 to 1993, and obtained Master's Degree (1989) and Ph. D Degree (1993) in Rock Mechanics and

Engineering Geology, then completed a post-doctoral work at the University of Paris XII (1993—1995). Mr. Liu had worked at Central South University of Technology (Central South University) from 1995 to 1997, and was promoted to Professor in 1996. From 1997 to 2004, Mr. Liu had accumulated some good working experiences both in CPI and IBM in USA. Then Mr. Liu came back to Central South University in 2004 and worked as Professor and Ph. D tutor. Since December 2013, Mr. Liu has been working at the College of Water Conservancy and Civil Engineering, South China Agricultural University.

Mr. Liu was awarded as outstanding young teacher for the year 1995 by the National Ministry of Education. He was also awarded by the State Education Commission Outstanding Overseas Ph. D Students Fund, IET and other incentives. Mr. Liu Chaired or undertook a number of provincial and ministerial fund research projects and enterprise scientific research projects, served as the team leader for a "973" sub-project once and the "973" project main researcher once, won the provincial and ministerial scientific and technological awards many times, including grand prize, first prize, second prize, etc. won also a second prize in the provincial teaching achievement awards. Mr. Liu has published 4 books and more than 80 research papers, and owns 1 national invention patent as well.

E-mail: alexliu@163.com

PREFACE 前言

世界上以英语作为官方语言的国家多达 171 个，占国家总数的 88.6%。在国际交流日益频繁和广泛深入的今天，作为世界上使用最广泛的语言——英语，其影响力早已席卷全球。英语的作用远远超出了作为一种语言在人民生活的方方面面所发挥的沟通交流作用，还关系到科学技术的革新、人类文明的进步以及世界的协同发展。在各行各业竞争越来越激烈的国际大环境下，具有较强的英语听说读写能力并能自如地将这种能力用于工作和生活之中，是决定一个人，尤其是高层次人才能否在事业和生活上更容易地获得较大成功的重要因素之一。因此，改变传统的教育理念，在人才培养过程中强调知识的活学活用，树立能力培养优先原则，发挥高等院校在人才培养上的优势，把年轻一代培养成具有扎实理论基础和系统专业知识的，具有国际视野和较强的使用专业外语进行学术交流沟通、自身能力培养和提高的卓越学习型人才，是新时期教育工作者义不容辞的职责。

随着改革开放的全方位深化与持续，高等院校教育教学改革也在加紧推进，培养国际化专业人才已经成为一种现实需求。在高校人才培养大纲中增设一定数量的双语专业课程，正在成为一种可以有效提高工科类学生专业英语听、说、读、写能力的重要手段，而且越来越受重视。事实上，在高校设置一定数量的双语教学课程既获得了教育管理部门的政策鼓励，也越来越受到广大师生的认可。然而，目前国内专业性较强的双语教材总量较少，既考虑专业知识的系统性，又兼顾科技英语学习技巧和专业英语能力培养的优秀双语教材就更少。

岩土工程类专业涵盖的范围较广，土木工程、岩土工程、水利水电工程、城市地下空间工程、建筑、交通、桥隧、港口码头等均与岩土工程关系密切。《岩土工程概论》是高校岩土工程类专业学生的重要专业基础课程。在专业课时被严重压缩的高校教学改革大环境下，将《岩土工程概论》设置成双语课程，既能教给学生有关岩土工程学科的系统专业基础知识，同时也能更有效地提高学生学习英语的兴趣，扩展学生专业词汇量，掌握常见科技英语表达模式，培养大学生专业科技英语的综合运用能力，为实现国际化人才的培养目标服务。基于该课程的教学改革项目"工程类本科生专业英语实际运用能力培养模式创新研究（GDJG20142093）"获 2014 年度广东省高等教育教学改革立项，为本教材的编写提供了支撑。

本书主编刘爱华教授长期从事岩土工程领域相关研究与教学工作，并获得过湖南省第十届高等教育教学成果二等奖，先后为本科生及研究生开设过《有限单元法及应用》《职业安全管理工程》及《岩土工程概论》等双语课程，能较好地把握专业双语课程在学习内容和形式上的特殊要求。华南农业大学水利与土木工程学院每年向 8 个土木工程专业班和 3 个水利水电工程专业班超过 300 名的学生开设了本课程，教学效果良好。

《岩土工程概论》（General Introduction to Geotechnical Engineering）以大学土木、水利、建筑、交通及其他岩土工程类本科学生为主要使用对象，兼顾研究生以及相关工程领域的科研与技术人员学习参考之需要，既考虑该课程专业知识上的系统性，同时兼顾科技英语学习的基本特点与要求，以中、英双语形式系统介绍岩土工程专业基础知识。本教材能帮助读者在学习岩土工程专业知识与技术的同时，在科技英语的听、说、读、写等综合能力上也得到同步锻炼，取得进步。

《岩土工程概论》（General Introduction to Geotechnical Engineering）在内容设置上进行了大胆的组合与优化。教材内容涵盖了岩土工程的基本分类方法与典型岩土工程，岩土材料的力学特性和测试方法，岩土工程相关计算理论、设计方法、施工技术和装备，岩土工程常见工程灾害与防治，数值化技术与岩土工程关系以及学科领域研究热点和前沿等方面内容。章与章之间在内容安排上既相互联系，又可自成一体，能方便满足不同层次读者的需求。

在书稿出版之际，特别感谢慧眼独具的中国水利水电出版社提供的合作平台与大力支持，感谢广州市水电建设工程有限公司为书稿出版给予的支持，感谢黄莉、欧阳帆、汪丽娜、闫晓满和邹家强等在书稿撰写过程中付出的辛勤劳动。双语书稿有其特殊性，可参照的经验和样本较少，加上作者水平所限，书中难免有疏漏之处，敬请广大读者批评指正。

作者

2016 年 5 月 31 日

CONTENTS 目录

前言

Chapter 1 Introduction to Geotechnical Engineering
岩土工程简介 ………………………………………………………… 1
1.1 Definition 定义 ………………………………………………… 3
1.2 History 发展历程 ……………………………………………… 7
1.3 General tasks and importance 基本任务与重要性 …………… 13

Chapter 2 Challenges in Geotechnical Engineering and Typical Geotechnical Engineering
岩土工程挑战与典型岩土工程 …………………………………… 19
2.1 Main challenges in geotechnical engineering 岩土工程主要挑战 …… 21
2.2 Human activities and geotechnical engineering 人类活动与岩土工程 …… 24
2.3 Typical kinds of geotechnical engineering 典型岩土工程 …… 29
2.4 Lateral earth support structures 侧向土支护结构 …………… 37
2.5 Earth structures 土工建筑物 …………………………………… 44

Chapter 3 Mechanical Properties of Soil/Rock and Testing
岩土力学特性与测试 ……………………………………………… 55
3.1 Definition of soil 土体定义 …………………………………… 57
3.2 Definition of rock 岩体定义 …………………………………… 61
3.3 Mechanical properties of soils 土的力学特性 ………………… 64
3.4 Mechanical properties of rocks 岩体的力学特性 ……………… 74
3.5 Testing methods and procedures 测试方法与流程 …………… 78

Chapter 4 Theoretical Calculation of Geotechnical Engineering Problems
岩土工程问题的理论计算 ………………………………………… 87
4.1 Definition of geotechnical engineering problems 岩土工程问题定义 …… 89
4.2 Theories to resolve geotechnical engineering problems 岩土工程问题求解理论 …… 98
4.3 Theoretical errors and geotechnical engineering needs
　　理论误差与岩土工程需要的关系 …………………………… 109

Chapter 5 Geotechnical Engineering Design
岩土工程设计 ……………………………………………………… 115
5.1 Rules to be followed 必须遵循的准则 ………………………… 117

5.2 General design methods 一般设计方法 ·················· 122

Chapter 6 Geotechnical Engineering Construction
岩土工程施工 ·················· 131
6.1 General construction methods 一般施工方法 ·················· 133
6.2 Construction equipments 施工设备 ·················· 151
6.3 Construction cost and safety 工程建设成本与安全性 ·················· 157

Chapter 7 Disasters in Geotechnical Engineering
岩土工程灾害 ·················· 171
7.1 Definition of engineering disasters 工程灾害定义 ·················· 173
7.2 Basic causes to engineering disasters 造成工程灾害的基本原因 ·················· 175
7.3 Geotechnical engineering disasters 岩土工程灾害 ·················· 181
7.4 Engineering disasters prevention 工程灾害防治 ·················· 194

Chapter 8 Numerical Technologies and Geotechnical Engineering
数值技术与岩土工程 ·················· 209
8.1 Basic concepts in numerical technologies 数值技术基本概念 ·················· 211
8.2 Numerical methods and their advantages 常见数值方法及其优越性比较 ·················· 214
8.3 Future challenges 学科领域未来的挑战 ·················· 222

Appended Pictures 附图 ·················· 225
References 参考文献 ·················· 230

Chapter 1
Introduction to Geotechnical Engineering

岩土工程简介

1.1 Definition 定义

【Text】

Everything you see around you is supported by soil or rock. Anything that is not supported by soil or rock, either floats, flies or falls down.

Civil engineering is a professional engineering discipline that deals with the design, construction, and maintenance of the physical and naturally built environment, including works like roads, bridges, canals, dams, and buildings. Civil engineering is the oldest engineering discipline after military engineering, and it was defined to distinguish non-military engineering from military engineering. It is traditionally broken into several sub-disciplines including environmental engineering, geotechnical engineering, geophysics, geodesy, control engineering, structural engineering, transportation engineering, earth science, atmospheric sciences, forensic engineering, municipal or urban engineering, water resources engineering, materials engineering, offshore engineering, quantity surveying, coastal engineering, surveying, and construction engineering. Civil engineering takes place on all levels: in the public sector from municipal through to national governments, and in the private sector from individual homeowners through to international companies.

Geotechnical engineering is obviously the branch of civil engineering concerned with the engineering behaviour of earth materials. In the other words, Geotechnical engineering is specialized in civil engineering field research about the engineering properties of soil and rock mass and its application.

Geotechnical engineering mainly includes the following main aspects: soil science, geology (including hydrology), engineering survey, foundation (foundation treatment, foundation engineering), tunnel excavation, foundation pit excavation, excavation engineering, supporting engineering (foundation pit supporting, slope supporting and debris flow control), engineering detection and monitoring etc. Actually, the above problems can be summed up in the three eternal classic problems in soil mechanics: slope stability, soil pressure and bearing capacity of foundation.

Geotechnical engineering is thus an area of civil engineering concerned with the rock and soil that support civil engineering systems. Knowledge from the fields of geology, material science and testing, mechanics, and hydraulics are applied by geotechnical engineers to safely and economically design foundations, retaining walls, and similar structures. Environmental concerns in relation to groundwater and waste disposal have spawned a new area of study called geo-environmental engineering where biology and chemistry are important. Some of the unique difficulties of geotechnical engineering are the result of the variability and properties of soil. Boundary conditions are often well defined in other branches of civil engineering, but with soil, clearly defining these conditions can be impossible. The material properties and behavior of soil are also difficult to predict due to the variability of soil and limited investigation. This contrasts with the relatively well defined material properties of steel and concrete used in other areas of civil engineering. Soil mechanics, which describes the behavior of soil, is also complicated because soils exhibit nonlinear (stress-dependent) strength, stiffness, and dilatancy (volume change associated with application of shear stress).

【Key words】

soil n. 泥土
rock n. 岩石
civil engineering 土木工程
geotechnical engineering 岩土工程
environmental engineering 环境工程
structural engineering 结构工程
offshore engineering 近海工程
coastal engineering 海岸工程
municipal or urban engineering 市政工程
geology n. 地质学
foundation n. 地基（基础）
excavation n. 开挖
supporting n. 支护
slope stability 边坡稳定
soil pressure 土压力
bearing capacity 承（载）量（能力）
groundwater n. 地下水
waste disposal 废物弃置（场）
variability n. 变化（性）
boundary condition 边界条件
investigation n. 调查
nonlinear adj. 非线性的
strength n. 强度
stress n. 应力
stiffness n. 刚度
dilatancy n. 膨胀性

【Translation】

Everything you see around you **is supported by** soil or rock. Anything that is not supported by soil or rock, either floats, flies or falls down.

我们周围能见到的所有东西都是为岩土所支撑的。不被岩土所支撑的东西要么浮在水面,要么飞在空中,要么坠落地下。

Civil engineering is a professional engineering discipline that deals with the design, construction, and maintenance of the physical and naturally built environment, including works like roads, bridges, canals, dams, and buildings. Civil engineering is the oldest engineering discipline after military engineering, and it was defined to distinguish non-military engineering from military engineering. **It is traditionally broken into** several sub-disciplines including environmental engineering, geotechnical engineering, geophysics, geodesy, control engineering, structural engineering, transportation engineering, earth science, atmospheric sciences, forensic engineering, municipal or urban engineering, water resources engineering, materials engineering, offshore engineering, quantity surveying, coastal engineering, surveying, and construction engineering. Civil engineering **takes place on all levels**: **in the public sector** from municipal through to national governments, **and in the private sector** from individual homeowners through to international companies.

土木工程是一门专门处理道路、桥梁、运河、水坝和建筑物等工程的设计、施工以及工程完成后周边环境维护等工作的工程学科。土木工程是继军用工程之后最为古老的工程学科,用以区分军用工程和非军用工程。传统上将土木工程细分为环境工程、岩土工程、地球物理、大地测量、控制工程、结构工程、交通工程、土壤学、大气科学、法律工程学、市政工程、水资源工程、材料工程、海洋工程、数量测量、海岸工程、测量、建筑工程等子学科。土木工程涵盖各个层面:在公共领域,可以从市政府到国家机构;在私营层面,则可以从私营业主到国际化企业。

Geotechnical engineering is obviously the branch of civil engineering concerned with the engineering behaviour of earth materials. In other words, geotechnical engineering **is specialized in** civil engineering **field research about** the engineering properties of soil and rock mass and its application.

显然,岩土工程是土木工程中与土工材料工程性质密切相关的一个分支。换句话说,岩土工程专注于土木工程领域关于岩土体工程性质研究及其应用。

Geotechnical engineering mainly includes the following main aspects: soil science, geology (including hydrology), engineering survey, foundation (foundation treatment, foundation engineering), tunnel excavation, foundation pit excavation, excavation engineering, supporting engineering (foundation pit supporting, slope supporting and debris flow control), engineering detection and monitoring etc. Actually, **the above problems can be summed up in** the three eternal classic problems in soil mechanics: slope stability, soil pressure and bearing capacity of foundation.

岩土工程主要包括以下几个方面：土壤学、地质（包括水文）、工程勘察、基础工程（地基处理，地基工程）、隧道开挖、基坑开挖、开挖工程、支护工程（基坑支护、边坡支护和泥石流治理）、工程检测和监测等。事实上，上述问题完全可以归结为土力学中最经典的三大问题：边坡稳定性、土压力和地基承载力。

Geotechnical engineering is thus an area of civil engineering concerned with the rock and soil that support civil engineering systems. Knowledge from the fields of geology, material science and testing, mechanics, and hydraulics are applied by geotechnical engineers to safely and economically design foundations, retaining walls, and similar structures. **Environmental concerns in relation to** groundwater and waste disposal **have spawned a new area of study called** geoenvironmental engineering **where** biology and chemistry are important. Some of the unique difficulties of geotechnical engineering are the result of the variability and properties of soil. Boundary conditions are often well defined in other branches of civil engineering, but with soil, clearly defining these conditions can be impossible. The material properties and behavior of soil are also difficult to predict due to the variability of soil and limited investigation. This contrasts with the relatively well defined material properties of steel and concrete used in other areas of civil engineering. Soil mechanics, which describes the behavior of soil, is also complicated because soils exhibit nonlinear (stress-dependent) strength, stiffness, and dilatancy (volume change associated with application of shear stress).

因此，岩土工程属于与岩土体密切相关的土木工程领域，并为土木工程体系提供支撑。岩土工程师们运用地质学、材料科学与测试技术、力学和水力学等方面的知识进行基础工程、挡土墙工程和其他类似构筑物的安全而又经济的设计。此外，与地下水和废物处理有关的环境问题催生了一个新的研究领域，称为环境岩土工程，生物和化学知识在该领域至关重要。岩土工程中的一些独特的难题源自土体的多样性及其性质。在土木工程的其他分支中，边界条件往往是很好定义的，但对土体而言，明确定义这些边界条件可能是行不通的。同样，由于土体的多样性以及有限的调查数据使得土体的材料性质和特性也难以进行准确预测。这一点与土木工程其他领域中用到的钢材和混凝土的材料性能能获得比较明确的定义相比较，差别明显。描述土体性质的土力学，也因为土体具有非线性强度（与应力相关）、刚度和剪胀性（体积随施加的剪应力而变化）而变得复杂。

Civil engineering is a branch of engineering that deals with the design and construction of structures that are intended to be stationary, such as buildings and houses, tunnels, bridges, canals, highways, airports, port facilities, and road beds for railroads.

土木工程是工程学的一个分支，涉及各类固定建筑物的设计和建造，包括大楼、房屋、隧道、桥梁、运河、高速公路、机场、港口设施以及铁路路基等。

Among its subdivisions are structural engineering, dealing with permanent structures; hydraulic engineering, dealing with the flow of water and other fluids; and environment/sanitary engineering, dealing with water supply, water purification, and sewer system, as well as urban planning and design.

土木工程的子学科又可分为结构工程，主要研究永久性建筑；水利工程，主要研究水和其他流体的输送；环境/卫生工程，主要研究水的供应、净化和排水系统，以及城市规划与设计等。

【Important sentences】

1. … is supported by…
   ……被……所支撑。
2. It is traditionally broken into…
   传统上，被细分为……
3. It takes place on all levels: in the public sector…, and in the private sector…
   它涵盖各个层面：在公共领域，……；在私营层面，……
4. … is specialized in… field research about…
   ……专注于……领域关于……
5. Actually, the above problems can be summed up in…
   事实上，上述问题完全可以归结为……
6. Environmental concerns in relation to… have spawned a new area of study called … where… are important.
   与……有关的环境问题催生了一个新的研究领域，称为……，在这个该领域，……是至关重要的。（where 引导状语从句）

1.2 History 发展历程

【Text】

Humans have historically used soil as a material for flood control, irrigation purposes, burial sites, building foundations, and as construction material for buildings. First activities were linked to irrigation and flood control, as demonstrated by traces of dykes, dams, and canals dating back to at least 2000 BCE that were found in ancient Egypt, ancient Mesopotamia and the Fertile Crescent, as well as around the early settlements of Mohenjo Daro and Harappa in the Indus valley. As the cities expanded, structures were erected supported by formalized foundations; Ancient Greeks notably constructed pad footings and strip-and-raft foundations. Until the 18th century, however, no theoretical basis for soil design had been developed and the discipline was more of an art than a science, relying on past experience.

Several foundation-related engineering problems, such as the Leaning Tower of Pisa, prompted scientists to begin taking a more scientific-based approach to examining the subsurface. The earliest advances occurred in the development of earth pressure theories for the construction of retaining walls. Henri Gautier, a French Royal Engineer, recognized the "natural slope" of different soils in 1717, an idea later known as the soil's angle of repose. A rudimentary soil classification system was also developed based on a material's unit weight, which is no longer considered a good indication of soil type.

The application of the principles of mechanics to soils was documented as early as 1773 when Charles Coulomb (a physicist, engineer, and army Captain) developed improved methods to determine the earth pressures against military ramparts. Coulomb observed that, at failure, a distinct slip plane would form behind a sliding retaining wall and he suggested that the maximum shear stress on the slip plane, for design purposes, was the sum of the soil cohesion, c, and friction $\sigma\tan\varphi$, where σ is the normal stress on the slip plane and φ is the friction angle of the soil. By combining Coulomb's theory with Christian Otto Mohr's 2D stress state, the theory became known as Mohr-Coulomb theory. Although it is now recognized that precise determination of cohesion is impossible because c is not a fundamental soil property, the Mohr-Coulomb theory is still used in practice today.

In the 19th century Henry Darcy developed what is now known as Darcy's Law describing the flow of fluids in porous media. Joseph Boussinesq (a mathematician and physicist) developed theories of stress distribution in elastic solids that proved useful for estimating stresses at depth in the ground; William Rankine, an engineer and physicist, developed an alternative to Coulomb's earth pressure theory. Albert Atterberg developed the clay consistency indices that are still used today for soil classification. Osborne Reynolds recognized in 1885 that shearing causes volumetric dilation of dense and contraction of loose granular materials.

Modern geotechnical engineering is said to have begun in 1925 with the publication of *Erdbaumechanik* by Karl Terzaghi (a civil engineer and geologist). Considered by many to be the father of modern soil mechanics and geotechnical engineering, Terzaghi developed the principle of effective stress, and demonstrated that the shear strength of soil is controlled by effective stress. Terzaghi also developed the framework for theories of bearing capacity of foundations, and the theory for prediction of the rate of settlement of clay layers due to consolidation. In his 1948 book, Donald Taylor recognized that interlocking and dilation of densely packed particles contributed to the peak strength of a soil. The interrelationships between volume change behavior (dilation, contraction, and consolidation) and shearing behavior were all connected via the theory of plasticity using critical state soil mechanics by Roscoe, Schofield, and Wroth with the publication of "On the Yielding of Soils" in 1958. Critical state soil mechanics is the basis for many contemporary advanced constitutive models describing the behavior of soil.

【Key words】
flood control 防洪
irrigation n. 灌溉
burial site 掩埋点
dyke n. 堤坝
dam n. 水坝
canal n. 运河
BCE = Before the Common Era 公元前
settlement n. 定居点
Egypt n. 埃及
Mesopotamia 美索不达米亚（亚洲西南部 Tigris 和 Euphrates 两河流域间的古王国，今伊拉克所在地）
Fertile Crescent "新月沃地"指中东两河流域及附近一连串肥沃的土地
Mohenjo Daro and Harappa in the Indus valley 在印度河流域的摩亨佐达罗和哈拉帕
formalize vt. 使正式，形式化 vi. 拘泥于形式
pad footing 垫片式基础
strip-and-raft foundation 带形和筏形基础
retaining wall 挡土墙
angle of repose 静止角
unit weight 单位重量
military rampart 军用防御挡墙
2D stress state 平面应力状态
cohesion n. 黏聚力
friction n. 摩擦
normal stress 法向应力
Mohr-Coulomb theory 莫尔-库仑理论
Darcy's Law 达西定律
clay consistency indices（index 的复数） 黏土一致性指数

effective stress 有效应力
Karl Terzaghi 卡尔·太沙基
framework n. 组织架构
dilation n. 膨胀
contraction n. 收缩
consolidation n. 固结，压实
plasticity n. 塑性
theory of plasticity 塑性理论
critical state soil mechanics 临界状态土力学
constitutive model 本构模型

【Translation】

Humans have historically used soil as a material for flood control, irrigation purposes, burial sites, building foundations, and as construction material for buildings. First activities were linked to irrigation and flood control, as demonstrated by traces of dykes, dams, and canals dating back to at least 2000 BCE that were found in ancient Egypt, ancient Mesopotamia and the Fertile Crescent, as well as around the early settlements of Mohenjo Daro and Harappa in the Indus valley. As the cities expanded, structures were erected and supported by formalized foundations; Ancient Greeks notably constructed pad footings and strip-and-raft foundations. Until the 18th century, however, no theoretical basis for soil design had been developed and the discipline was more of an art than a science, relying on past experience.

历史上，人类已经有使用土体作为防洪、灌溉、墓地、建筑地基和楼房施工材料的记载。与灌溉和防洪有关的早期活动至少可以追溯到公元前 2000 年，这一点可以通过在古埃及、古美索不达米亚和新月沃地，以及在印度河流域的摩亨佐达罗和哈拉帕的早期聚集地周边所发现的堤坝、水坝和运河等得到印证。随着城市的扩张，构筑物建立在规范化的基础之上；例如古希腊人就建造了垫片式地基以及带形和筏形基础。然而，直到 18 世纪，还没有形成有关土体设计的理论基础，那时的土体设计学科更像是一门依靠过往经验的手艺而不是科学。

Several foundation-related engineering **problems, such as** the Leaning Tower of Pisa, **prompted scientists to begin taking a more scientific-based approach to examining the subsurface.** The earliest advances occurred in the development of earth pressure theories for the construction of retaining walls. Henri Gautier, a French Royal Engineer, recognized the "natural slope" of different soils in 1717, an idea later known as the soil's angle of repose. A rudimentary soil classification system was also developed based on a material's unit weight, which is no longer considered a good indication of soil type.

一些与基础有关的工程问题，例如比萨斜塔，促使科学家们在检查地下情况时开始采取更科学的方法。在关于挡土墙建设的土压力理论的发展形成过程中出现了最早期的一些进步。法国皇家工程师亨利·戈蒂埃，在1717年认识到不同类型土的"自然边坡"问题，后来，这一想法被称作土的静止角。同时，还发展形成了一种基于材料单位重量的土体基本分类体系，而材料的单位重量不再被认为是一个好的用于区分土体类型的指标。

The application of the principles of mechanics to soils was documented as early as 1773 when Charles Coulomb (a physicist, engineer, and army Captain) developed improved methods to determine the earth pressures against military ramparts. Coulomb observed that, at failure, a distinct slip plane would form behind a sliding retaining wall and he suggested that the maximum shear stress on the slip plane, for design purposes, was the sum of the soil cohesion, c, and friction $\sigma\tan\varphi$, where σ is the normal stress on the slip plane and φ is the friction angle of the soil. By combining Coulomb's **theory with** Christian Otto Mohr's 2D stress state, **the theory became known as** Mohr-Coulomb theory. Although it is now recognized that precise determination of cohesion is impossible because c is not a fundamental soil property, the Mohr-Coulomb theory is still used in practice today.

早在1773年就有了一些关于力学原理在土力学方面应用的记录，当时查尔斯·库仑（物理学家、工程师、陆军上尉）研究发展了一种改进过的用于确定军事防御工事上土压力的方法。库仑观察到，破坏时在移动的挡土墙后面会形成一个明显的滑移面，因此他建议，在设计时应考虑在滑动平面的最大剪应力，其大小是土壤黏聚力 c 和摩擦力 $\sigma\tan\varphi$ 的总和，其中 σ 是滑动面上的法向应力，φ 是土壤的摩擦角。将库仑理论与克里斯蒂安·奥托·莫尔的二维应力状态结合起来，就形成了著名的莫尔-库仑理论。尽管现在认识到黏聚力 c 的精确确定是不可能的，因为它不是土的一个基本属性，但今天莫尔-库仑理论依旧在实际工程中使用。

In the 19th century Henry Darcy developed what is now known as Darcy's Law describing the flow of fluids in porous media. Joseph Boussinesq (a mathematician and physicist) developed theories of stress distribution in elastic solids that proved useful for estimating stresses at depth in the ground; William Rankine, an engineer and physicist, developed an alternative to Coulomb's earth pressure theory. Albert Atterberg developed the clay consistency indices that are still used today for soil classification. Osborne Reynolds recognized in 1885 that shearing causes volumetric dilation of dense and contraction of loose granular materials.

19世纪，亨利·达西研究提出了现在众所周知的、用来描述孔隙介质中流体流动的达西定律。约瑟夫·布辛尼斯克（数学家和物理学家）提出了弹性固体中应力分布的理论，该理论被证实在估算地下深处的应力分布方面是有效的；威廉·朗肯（工程师和物理学家）提出了一种能替代库仑土压力理论的理论。艾伯特·阿太伯格提出了今天仍然被用于土壤分类的黏土一致性指数。奥斯鲍恩·雷诺在 1885 年指出，剪切能引起致密材料的体积膨胀以及松散颗粒材料的收缩。

Modern geotechnical engineering is said to have begun in 1925 with the publication of *Erdbaumechanik* by Karl Terzaghi (a civil engineer and geologist). **Considered by many to be the father of** modern soil mechanics and geotechnical engineering, Terzaghi **developed the principle of** effective stress, **and demonstrated that** the shear strength of soil is controlled by effective stress. Terzaghi also developed the framework for theories of bearing capacity of foundations, and the theory for prediction of the rate of settlement of clay layers due to consolidation. In his 1948 book, Donald Taylor recognized that interlocking and dilation of densely packed particles contributed to the peak strength of a soil. The interrelationships between volume change behavior (dilation, contraction, and consolidation) and shearing behavior were all connected via the theory of plasticity using critical state soil mechanics by Roscoe, Schofield, and Wroth with the publication of *On the Yielding of Soils* in 1958. Critical state soil mechanics is the basis for many contemporary advanced constitutive models describing the behavior of soil.

现代岩土工程据说始于 1925 年由卡尔·太沙基（土木工程师和地质学家）主办的《土力学》出版物。太沙基被许多人认为是现代土力学与岩土工程之父，他提出了有效应力原理，并论证了土的抗剪强度是由有效应力控制的。太沙基还构建了地基承载力理论以及黏土层因固结而沉降的速率预测理论的框架。唐纳德·泰勒在他 1948 年出版的书中指出，密集颗粒的相互咬合和膨胀影响土的峰值强度。体积变形特性（膨胀、收缩和固结）和剪切特性之间的相互关系均能通过罗斯科、斯科菲尔德和罗思 1958 年出版的《土的屈服》一书中基于临界状态土力学的塑性理论来建立联系。临界状态土力学是许多现代先进的描述土体特性的本构模型的基础。

【Important sentences】

1. Humans have historically used… as a material for…
   历史上，人类已经有使用……作为……材料的记载。
2. Several… problems, such as…, prompted scientists to begin taking a more scientific-based approach to examining the subsurface.
   一些与……有关的工程问题，例如……，促使科学家们在检查地下情况时开始采取一种更科学的方法。
3. By combining… theory with…, the theory became known as… theory.
   将……理论与……结合起来，就形成了著名的……理论。
4. Considered by many to be the father of…, sb. developed the principle of…, and demonstrated that…

某人被许多人认为是……之父，他提出了……原理，并论证了……（and 表示并列，demonstrated 后面的 that 引导宾语从句）

1.3 General tasks and importance 基本任务与重要性

【Text】

The importance of geotechnical engineering can hardly be overstated: buildings must be connected to the ground. Geotechnical engineering is concerned with soil properties, foundation, footings, soil-structure interaction and soil dynamics.

Geotechnical engineering uses principles of soil mechanics and rock mechanics to investigate subsurface conditions and materials; determine the relevant physical/mechanical and chemical properties of these materials; evaluate stability of natural slopes and man-made soil deposits; assess risks posed by site conditions; design earthworks and structure foundations; and monitor site conditions, earthwork and foundation construction.

A typical geotechnical engineering project begins with a review of project needs to define the required material properties. Then follows a site investigation of soil, rock, fault distribution and bedrock properties on and below an area of interest to determine their engineering properties including how they will interact with, on or in a proposed construction. Site investigations are needed to gain an understanding of the area in or on which the engineering will take place. Investigations can include the assessment of the risk to humans, property and the environment from natural hazards such as earthquakes, landslides, sinkholes, soil liquefaction, debris flows and rock falls.

A geotechnical engineer then determines and designs the type of foundations, earthworks, and/or pavement sub grades required for the intended man-made structures to be built. Foundations are designed and constructed for structures of various sizes such as high-rise buildings, bridges, medium to large commercial buildings, and smaller structures where the soil conditions do not allow code-based design.

Foundations built for above-ground structures include shallow and deep foundations. Retaining structures include earth-filled dams and retaining walls. Earthworks include embankments, tunnels, dikes, levees, channels, reservoirs, deposition of hazardous waste and sanitary landfills.

Geotechnical engineering is also related to coastal and ocean engineering. Coastal engineering can involve the design and construction of wharves, marinas, and jetties. Ocean engineering can involve foundation and anchor systems for offshore structures such as oil platforms.

The fields of geotechnical engineering and engineering geology are closely related, and have large areas of overlap. However, the field of geotechnical engineering is a specialty of engineering, where the field of engineering geology is a specialty of geology.

【Key words】
overstate *vt*. 把……讲得过分
soil-structure interaction　土体与结构间的相互作用
soil dynamics　土动力学
subsurface *adj*. 地表下的
relevant *adj*. 有关的，相关联的
natural slope　天然坡体
man-made *adj*. 人工的，人造的
earthwork *n*. 土方工程，土木工事
embankment *n*.（道路的）路堤，（河流的）岸堤
tunnel *n*. 隧道
dike *n*. 排水沟
levee *n*. 堤（沿岸构建，不同于拦水用的坝）
channel *n*. 运河，频道，海峡
reservoir *n*. 蓄水池，水库
deposition of hazardous waste　危险废物堆积场
sanitary landfills　卫生（废物）填埋场
assess *vt*. 评定；估价；评估
evaluate *vt*. 评价；对……评价　*vi*. 评价，估价
site investigation　现场（实地）调查
fault distribution　裂隙分布
bedrock *n*. 基岩
natural hazards　自然灾害
earthquake *n*. 地震
landslide *n*. 滑坡
sinkhole *n*. 塌陷（坑）
soil liquefaction　土壤液化
debris flows　泥石流
rock fall　岩石塌落
pavement *n*. 人行道；硬路面
sub grade　子（亚）等级，对应等级
high-rise building　高层建筑

code-based design　依法设计
coastal engineering　海岸工程
wharves *n*.　码头
marina *n*.　小艇船坞
jetty *n*.　防波堤
ocean engineering　海洋工程
offshore *adj*. 离开海岸的　*adv*.（指风）向海地，离岸地
offshore structure　离岸构筑物
oil platform　石油平台
engineering geology　工程地质

【Translation】

The importance of geotechnical engineering **can hardly be overstated**：buildings must be connected to the ground. Geotechnical engineering is concerned with soil properties, foundation, footings, soil-structure interaction and soil dynamics.

岩土工程的重要性怎么讲都不过分：建筑物离不开地基。岩土工程关注土的性质、基础、地基、土体与结构的相互作用和土动力学。

Geotechnical engineering uses principles of soil mechanics and rock mechanics to **investigate** subsurface conditions and materials；**determine** the relevant physical/mechanical and chemical properties of these materials；**evaluate** stability of natural slopes and man-made soil deposits；**assess risks** posed by site conditions；**design** earthworks and structure foundations；**and monitor** site conditions, earthwork and foundation construction.

岩土工程采用土力学和岩石力学原理，调查地下条件和材料，确定这些材料相关的物理或力学和化学性质，计算自然边坡和人工排土场的稳定性，评估现场条件带来的风险，设计土方工程和结构基础，监控现场条件、土方工程及基础施工。

A typical geotechnical engineering project begins with a review of project needs to define the required material properties. Then follows a site investigation of soil, rock, fault distribution and bedrock properties on and below an area of interest to determine their engineering properties including how they will interact with, on or in a proposed construction. Site investigations are needed to gain an understanding of the area in or on which the engineering will take place. Investigations can include the assessment of the risk to humans, property and the environment from natural hazards such as earthquakes, landslides, sinkholes, soil liquefaction, debris flows and rock falls.

一个典型的岩土工程项目开始于对项目需求的审查，从而决定所需的材料性能。然后，针对有意向的区域展开地上地下有关土体、岩体、断层分布以及基岩特性等的现场调查，以便确定其工程特性，弄清它们将如何与提出的施工建设产生相互作用。为了获得对将要进行工程施工的区域全面的了解需要开展现场调查。调查可能包括对由于诸如地震、滑坡、塌陷（坑）、土壤液化、泥石流和岩石塌落等自然灾害对人类、财产和环境的风险评估。

A geotechnical engineer then determines and designs the type of foundations, earthworks, and/or pavement sub grades required for the intended man-made structures to be built. Foundations are designed and constructed for structures of various sizes such as high-rise buildings, bridges, medium to large commercial buildings, and smaller structures where the soil conditions do not allow code-based design.

然后，岩土工程师就要针对有意要建设的人工建筑所需的基础类型、土方工程和/或路面等级等进行确定和设计。对于像高层建筑、桥梁、中型到大型商业建筑等各种尺寸的构筑物以及对于那些较小但土壤条件不符合规范要求设计的构筑物，都要进行基础设计与建造。

Foundations built for above-ground structures include shallow and deep foundations. Retaining structures include earth-filled dams and retaining walls. Earthworks include embankments, tunnels, dikes, levees, channels, reservoirs, deposition of hazardous waste and sanitary landfills.

为地面构筑物建造的基础包括浅基础和深基础。围护结构包括填土坝和挡土墙。土方工程包括路基、隧道、堤坝、防洪堤、渠道、水库、危险废物堆放场和垃圾填埋场。

Geotechnical engineering is also related to coastal and ocean engineering. Coastal engineering can involve the design and construction of wharves, marinas, and jetties. Ocean engineering can involve foundation and anchor systems for offshore structures such as oil platforms.

岩土工程也涉及海岸和海洋工程。海岸工程可以包括码头、小艇船坞、防波堤的设计和建造。海洋工程可以包括用于如石油平台一样的离岸构筑物的基础和锚固系统。

The fields of geotechnical engineering and engineering geology **are closely related，and have large areas of overlap. However，the field of** geotechnical engineering **is a specialty of** engineering，**where the field of** engineering geology **is a specialty of** geology.

岩土工程与工程地质领域关系密切,而且有许多重叠。然而,岩土工程领域属于工程学专项,而工程地质领域则属于地质学专项。

【Important sentences】

1. The importance of … can hardly be overstated.
   ……的重要性怎么讲都不过分。
2. Geotechnical engineering uses principles of soil mechanics and rock mechanics to investigate …; determine …; evaluate …; assess risks …; design …; and monitor …
   岩土工程采用土力学和岩石力学原理,调查……,确定……,计算……,评估……风险,设计……,监控……。(investigate、determine、evaluate、assess risks、design、monitor 几个动词均为并列动词,用分号连接)
3. The fields of A and B are closely related, and have large areas of overlap. However, the field of A is a specialty of C, where the field of B is a specialty of D.
   A 与 B 关系密切,而且有许多重叠。然而,A 领域属于 C 专项,而 B 领域则属于 D 专项。(However 表示转折,A 与 B 虽有很多重叠,却分别属于不同的领域专项)

Chapter 2
Challenges in Geotechnical Engineering and Typical Geotechnical Engineering
岩土工程挑战与典型岩土工程

2.1 Main challenges in geotechnical engineering
岩土工程主要挑战

【Text】

Geotechnical engineering is the oldest field in civil engineering, yet it is also the latest. Large-scale historical structures, such as the pyramids of the Egypt, bear witness to the astonishing knowledge in geotechnical engineering of the time. However, as with many other ancient crafts and techniques, their knowledge was lost. Thousands of years after pyramids, the engineers of this century are challenged with the structural integrity and bearing capacity of high-rise buildings, immense bridges, and huge dams in order to avoid sinking or collapse. Unlike the rare sighting of large structures in the ancient world, today, like civilization and commerce, they can be seen in all corners of the world. Bridges span rivers and gorges, while roads wind through snowy peaks or tunnel through the mountains in order to connect people to one another. High-rise buildings accommodate the needs of people in densely populated areas. The fate of these structures depends heavily on geotechnical engineers, whose foundation allow bridges and high-rise buildings to stand firm, whose sound judgment makes mountain roads safe even in bad weather, and whose precise analysis ensures the safety of tunnels. The soil and groundwater pollution has become more and more commonplace in today's highly industrial environment. It is absolutely necessary for the engineers to have a deep understanding of their construction materials—the on-site soil and rock conditions. Unlike the man-made materials of steel and concrete, it is much more difficult to grasp the characteristics of natural materials, since soil and rock conditions vary greatly all around the world. Thus, a geotechnical engineer's priority must be to understand the subtle similarities and differences in soil and rock properties, evaluate the engineering properties of the soil and rock on the site, and determine the suitable design and construction method. The uncertainties of natural elements and different geological properties pose quite a challenge to geotechnical engineers everywhere.

Civil engineering offers a particular challenge because almost every structure or system that is designed and built by civil engineers is unique. One structure rarely duplicates another exactly. Even when structures seem to be identical, site requirement or other factors generally result in modification. Large structures like dams, bridges, or tunnels may differ substantially from previous structures. Engineers are required to have solid knowledge of mathematics, physics, and chemistry. Mathematics is very important in all branches of engineering, so it is greatly stressed.

A civil engineer is a member of the civil engineering profession. They may work in research design, construction supervision, maintenance, or even in sales or management. Each of these areas involves different duties, different emphases, and different uses of the engineer's knowledge and experience.

【Key words】

astonishing *adj*. 惊人的
crafts and techniques   工艺与技术
integrity *n*. 完整
sinking or collapse   沉陷与崩塌
accommodate *vt*. 容纳；使适应；向……提供住处
on-site *adj*. 现场的；实地的
grasp *vt*. 抓住；了解
subtle *adj*. 微妙的；敏感的
design and construction method   设计与施工方法
duplicate *vt*. 复制；重复
substantially *adv*. 本质上，实质上；大体上；充分地；相当多地

【Translation】

Geotechnical engineering is the oldest field in civil engineering, yet it is also the latest. Large-scale historical structures, such as the pyramids of the Egypt, bear witness to the astonishing knowledge in geotechnical engineering of the time. However, as with many other ancient crafts and techniques, their knowledge was lost. Thousands of years after pyramids, the engineers of this century are challenged with the structural integrity and bearing capacity of high-rise buildings, immense bridges, and huge dams in order to avoid sinking or collapse. Unlike the rare sighting of large structures in the ancient world, today, like civilization and commerce, they can be seen in all corners of the world. Bridges span rivers and gorges, while roads wind through snowy peaks or tunnel through the mountains in order to connect people to one another. High-rise buildings accommodate the needs of people in densely populated areas. The fate of these structures **depends heavily on** geotechnical engineers, **whose** foundation allow bridges and high-rise buildings to stand firm, **whose** sound judgment makes mountain roads safe even in bad weather, **and whose** precise analysis ensures the safety of tunnels. The soil and groundwater pollution has **become more and more commonplace** in today's highly industrial environment. **It is absolutely necessary for** the engineers **to** have a deep understanding of their construction materials—the on-site soil and rock conditions. Unlike the man-made materials of steel and concrete, it is much more difficult to grasp the characteristics of natural materials, since soil and rock conditions vary greatly all around the world. Thus, a geotechnical engineer's priority must be to understand the subtle similarities and differences in soil and rock properties, evaluate the engineering properties of the soil and rock on the site, and determine the suitable design and construction method. The uncertainties of natural elements and different geological properties pose quite a challenge to geotechnical engineers everywhere.

岩土工程是土木工程中最久远的一个分支,但又是最前沿的。大规模的历史构筑物,如埃及的金字塔,展示了当时不可思议的岩土工程知识。然而,和许多其他古代工艺和技术一样,那些知识失传了。金字塔建成后已过数千年之久,本世纪的工程师依旧面临为了避免高层建筑、大桥和大坝的沉降或坍塌而必须保证其结构完整性和承载能力的挑战。不同于古时大型建筑物难得一见,今天,就像文明和商贸一样,大型建筑物遍及世界的每个角落。飞跨江、谷两岸的桥梁,穿山越岭的道路使人民能够彼此连通。高层建筑满足了人口密集区人们的需求。这些构筑物的命运严重依赖于岩土工程师,他们的理论基础使得大桥和高层建筑屹立不倒,他们准确的判断使得山中道路即便在恶劣天气中安全畅通,他们的精确分析确保隧道安全。今天,在高度工业化的环境下,土壤和地下水的污染问题已经变得越来越常见。工程师对建筑材料,即现场岩土的条件有深入细致的了解是绝对必要的。不同于钢筋和水泥这样的人工材料,掌握这些天然材料的性能要难得多,因为世界各地的岩土性质相差悬殊。因此,岩土工程师的首要任务是必须弄清岩土性质上的细微异同,评判现场岩土的工程特性,确定适当的设计和施工方法。自然因素的不确定性以及不一样的地质特性给世界各地的岩土工程师带来了不小的挑战。

Civil engineering offers a particular challenge because almost every structure or system that is designed and built by civil engineers is unique. One structure rarely duplicates another exactly. Even when structures seem to be identical, site requirement or other factors generally result in modification. Large structures like dams, bridges, or tunnels may differ substantially from previous structures. Engineers are required to have solid knowledge of mathematics, physics, and chemistry. Mathematics is very important in all branches of engineering, so it is greatly stressed.

土木工程学科提出了一种特别的挑战,因为几乎每一幢由土木工程师设计和建造的构筑物或体系都是独一无二的。一种构筑物很少会被另一种构筑物完全复制。即便当建筑物看似相同,场地要求或其他因素常常会带来改变。大型构筑物如堤坝、桥梁或隧道工程可能与以前的构筑物有实质性差异。要求工程师具有坚实的数学、物理和化学知识。在工程科学的各个学科中,数学是非常重要的,这点要予以重点强调。

A civil engineer is a member of the civil engineering profession. They may work in research design, construction supervision, maintenance, or even in sales or management. Each of these areas involves different duties, different emphases, and different uses of the engineer's knowledge and experience.

Chapter 2 Challenges in Geotechnical Engineering and Typical Geotechnical Engineering 岩土工程挑战与典型岩土工程

土木工程师是土木工程专业队伍中的一员。他们可以从事研究设计、施工监理、维护维修、甚至销售或管理等工作。这些领域的每项工作有不同的职责，不同的侧重点，要运用不同的工程知识和经验。

【Important sentences】

1. The fate of these structures depends heavily on geotechnical engineers, whose foundation allow bridges and high-rise buildings to stand firm, whose sound judgment makes mountain roads safe even in bad weather, and whose precise analysis ensures the safety of tunnels.
   depends heavily on 严重依赖于，heavily 作 depends on 的状语。whose…, whose…, and whose…作 engineers 的并列定语从句。
2. … become more and more commonplace.
   （固定句型）变得越来越常见。
3. It is absolutely necessary for sb. to do sth.
   某人做某事是绝对必要的。

2.2 Human activities and geotechnical engineering
人类活动与岩土工程

【Text】

What is civil engineering? Engineering is the practical application of the findings of theoretical science so that they can be put to work for the benefit of mankind. Engineer is one of the oldest occupations in the history of mankind. Without the skills included in the field of engineering, our present-day civilization could never have evolved.

Civil engineering is a subject that has a very close relationship with human. It has a long history, but is also full of vigor.

Now civil engineering is defined as the engineering that is the floorboard of technology to build all kinds of engineering facilities. It means that the civil engineering can apply to the technology activities such as materials, equipment and the survey, design, construction, maintenance etc., and also refer to the object of construction, namely to the engineering facilities that have been serving our life, e.g. the productions directly or indirectly built on the ground or underground, onshore or underwater, such as houses, road, railways, bridges, dams, port, power stations, airports, hydraulic structures, water supply and drainage and protection engineering and so on.

It is the primary purpose and starting point of the civil engineering to provide good function, comfortable and beautiful space and channel to meet human activities' need, and excellent performance for waterproof infestations, water conservancy and environmental management.

Shelter, one of the primary needs of mankind, is provided by civil engineers. The efficient planning of water supply and irrigation systems increases the food production in a country. Shelters, apart from just being shelters, have been constructed by civil engineers to provide a peaceful and comfortable life.

The development of civil engineering went through a long phylogeny. Building house was the earliest civil engineering projects which mankind began, in Paleolithic times. People perched on the trees, or lived in a natural cave, afterwards, as the population increased, people started to use stone or branches to build man-made shelters by intimating natural cave. This was the beginning of civil engineering. From slave society, feudal society to today, various aspects in civil engineering, such as material, construction technology and theory have made great progress. The engineering marvels of the world, starting from the pyramids to today's thin shell structures, are the result of the development in civil engineering.

Any discipline of engineering is a vast field with various specializations. The major specializations of civil engineering are structural engineering, geotechnical engineering, fluid mechanics, hydraulics and hydraulic machines, transportation engineering, water supply, sanitary and environmental engineering, irrigation engineering, surveying, leveling and remote sensing, and so on.

Civil engineering material is the foundation of civil engineering. As the new material appears, it will promote innovation in architectural forms, thus the design and construction process will correspondingly be improved and innovated. Civil engineering material is gradually developed from scratch associated with human society advances and productive forces development of society, including several stages such as the stage from natural material to artificial material, the stage from handicraft production to industrial production and so on. In the Neolithic age, humans began using the natural material such as clay wood to build house. With the development of human production tools, natural stone material began to be used extensively. In the 19th century, the new civil engineering materials i.e. steel, cement, concrete and reinforced concrete appeared and were applied. The revolution of civil engineering materials arose in the history. In the 20th century, plastic alloy, stainless steel and some new civil engineering materials with special function appeared, which made civil engineering materials in the 21st century develop towards light weight, high strength, high durability, functionality, intelligent high-performance materials and environment-friendly green building materials.

In today's world, civil engineering works are all over the world, such as famous Sydney Opera House, Shanghai World Financial Center, and so on. The development is exciting both in the subject and its application. As the urban construction and road construction continues to heat up, the employment situation of civil engineering may certainly continue to rise in recent years as well.

【Key words】

mankind n. 人类
present-day adj. 现今的，目前的
vigor adj. 精力；活力
floorboard n. 地板，基础
engineering facilities 工程设施
hydraulic structures 水利设施
water supply and drainage 给排水
waterproof adj. 不透水的；防水的
infestation n. 蔓延；侵扰
phylogeny n. 发展史
paleolithic adj. ［考古］旧石器时代的
marvel n. 奇迹
improvement and innovation 改良，创新
neolithic adj. 新石器时代的
environment-friendly adj. 环境友好型的

【Translation】

What is civil engineering? Engineering is the practical application of the findings of theoretical science so that they can be put to work for the benefit of mankind. Engineer is one of the oldest occupations in the history of mankind. **Without** the skills included in the field of engineering, our present-day civilization **could never have** evolved.

究竟什么是土木工程？工程是理论科学研究成果的实际应用，因而可以造福人类。工程师则是人类历史上最古老的职业之一。要是没有工程领域累积的各项技能，我们当今的文明可能永远都不会发展进步。

Civil engineering is a subject that has a very close relationship with human. It has a long history, but is also full of vigor.

土木工程是跟人类有着十分密切联系的学科，该学科拥有悠久的历史，而且一

直都保持着勃勃生机。

Now civil engineering **is defined as** the engineering that is the floorboard of technology to build all kinds of engineering facilities. It means that the civil engineering can apply to the technology activities such as materials, equipment and the survey, design, construction, maintenance etc., and also refer to the object of construction, namely to the engineering facilities that have been serving our life, e. g. the productions directly or indirectly built on the ground or underground, onshore or underwater, such as houses, road, railways, bridges, dams, port, power stations, airports, hydraulic structures, water supply and drainage and protection engineering and so on.

现在，土木工程被定义为这样一种工程，即建造各类工程设施的工程技术的总和。这意味着土木工程既能适用于材料、装备和勘察、设计、施工、维护等技术性活动，也关系到具体的施工主体，即那些为我们生活服务的工程设施，如那些直接或间接建造于地面或地下、岸边或水下的各种工程，比如房屋、公路、铁路、桥梁、水坝、港口、电站、机场、水利设施、供水和排水及防护工程等。

It's the primary purpose and starting point of the civil engineering to provide good function, comfortable and beautiful space and channel to meet human activities' need, and excellent performance for waterproof infestations, water conservancy and environmental management.

为满足人类活动的需要，提供功能良好、舒适、美观的空间和通道，兼具防水患、兴水利和利环境的优良作用，这正是土木工程的主要目的和出发点。

Shelter, one of the primary needs of mankind, is provided by civil engineers. The efficient planning of water supply and irrigation systems increases the food production in a country. Shelters, apart from just being shelters, have been constructed by civil engineers to provide a peaceful and comfortable life.

作为人类基本需求之一的庇护所（住房）是由土木工程师提供的。供水和灌溉系统的有效规划可以提高国家的粮食产量。由土木工程师建造的各类庇护所，除了发挥其作为居所的作用以外，也为人们提供一个和平舒适的生活环境。

The development of civil engineering went through a long phylogeny. Building house was the earliest civil engineering projects which mankind began, in Paleolithic times. People perched on the trees, or lived in a natural cave, afterwards, as the population increased, people started to use stone or branches to build man-made shelters by imitating natural cave. This was the beginning of civil engineering. From slave society, feudal society to today, various aspects in civil engineering, such as material, construction technology and theory have made great progress. The engineering marvels of the world, starting from the pyramids to today's thin shell structures, are the result of the development in civil engineering.

土木工程的发展经历了一个漫长的发展过程。在旧石器时代，盖房子是人类开始的最早的土木工程项目。人们栖息在树上或生活在天然洞穴里，后来随着人口的增加，人们开始使用石头或树枝，通过模仿自然洞室建造人工庇护所，这就是土木工程的开始。从奴隶社会、封建社会到今天，土木工程各个领域，如材料、施工技术和理论等都取得了巨大进展。全世界的工程奇迹，从金字塔到今天的薄壳结构，都是土木工程发展的结果。

Any discipline of engineering is a vast field with various specializations. **The major specializations** of civil engineering **are** structural engineering, geotechnical engineering, fluid mechanics, hydraulics and hydraulic machines, transportation engineering, water supply, sanitary and environmental engineering, irrigation engineering, surveying, leveling and remote sensing and so on.

任何工程学科都是一个庞大的涵盖多种专业的领域。土木工程的主要专业有结构工程、岩土工程、流体力学、水力学和水利机械、交通运输工程、供水管路、卫生和环境工程、灌溉工程、测量学、水准测量和遥感等。

Civil engineering material is the foundation of civil engineering. As the new material appears, it will promote innovation in architectural forms, thus the design and construction process will correspondingly be improved and innovated. Civil engineering material is gradually developed from scratch associated with human society advances and productive forces development of society, including several stages such as the stage from natural material to artificial material, the stage from handicraft production to industrial production and so on. In the Neolithic age, humans began using the natural material such as clay wood to build house, with the development of human production tools, natural stone material began to be used extensively. In the 19th century, the new civil engineering materials i. e. steel, cement, concrete and reinforced concrete appeared and were applied. The revolution of civil engineering materials arose in the history. In the 20th century, plastic alloy, stainless steel and some new civil engineering materials with special function appeared, which made civil engineering materials in the 21st century develop towards light weight, high strength, high durability, functionality, intelligent high-performance materials and environment-friendly green building materials.

土木工程材料是土木工程的基础，新材料的出现将促进建筑形式的革新，相应地带来设计和施工过程的改进和创新。土木工程材料随着人类社会进步和社会生产力的发展从零开始逐步发展而来，包括从自然材料到人工材料、从手工制作到工业生产等几个阶段。在新石器时代，人类开始使用天然的材料，比如用黏土、木头造房子，随着人类生产工具的发展，天然石材开始被广泛使用，19世纪，出现了钢材、水泥、混凝土和钢筋混凝土等土木工程材料并获得应用，造就了历史上土木工程材料发生革命的时代。20世纪出现了塑料合金、不锈钢和一些具有特殊功能的新型土木工程材料，这使得21世纪的土木工程材料向着轻质高强、强耐久性、功能性、智能化的高性能材料和环保型绿色建筑材料发展。

In today's world, civil engineering works are all over the world, such as famous Sydney Opera House, Shanghai World Financial Center, and so on. The development is exciting both in the subject and its application. As the urban construction and road construction continues to heat up, the application situation of civil engineering may certainly continue to rise in recent years as well.

在当今世界，土木工程结构遍布世界各地，比如著名的悉尼歌剧院、上海世贸中心等。土木工程专业的发展无论是在学科本身，还是在其应用方面都令人感到兴奋。随着城市建设和公路建设的继续升温，土木工程的应用近些年还将持续上升。

【Important sentences】

1. Without the skills included in the field of engineering, our present-day civilization could never have evolved.

    Without…could never have done 表示虚拟语气，"没有……，就不会有……"。
2. Be defined as…

    被定义为……
3. The major specializations of civil engineering are…

    土木工程的主要专业有……

2.3　Typical kinds of geotechnical engineering　典型岩土工程

【Text】

Typical kinds of geotechnical engineering include foundations, lateral earth support structures, earth structures, slope stabilization structures, marine geotechnical engineering structures, geosynthetics products etc.

Foundations

A building's foundation transmits loads from buildings and other structures to the earth. Geotechnical engineers design foundations based on the load characteristics of the structure and the properties of the soils and/or bedrock at the site. In general, geotechnical engineers: ① Estimate the magnitude and location of the loads to be supported; ② Develop an investigation plan to explore the subsurface; ③ Determine necessary soil parameters through field and lab testing (e. g., consolidation test, triaxial shear test, vane shear test, standard penetration test); ④ Design the foundation in the safest and most economical manner.

The primary considerations for foundation support are bearing capacity, settlement, and ground movement beneath the foundations. Bearing capacity is the ability of the site soils to support the loads imposed by buildings or structures. Settlement occurs under all foundations in all soil conditions, though lightly loaded structures or rock sites may experience negligible settlements. For heavier structures or softer sites, both overall settlement relative to unbuilt areas or neighboring buildings, and differential settlement under a single structure, can be concerns. Of particular concern is settlement which occurs over time, as immediate settlement can usually be compensated for during construction. Ground movement beneath a structure's foundations can occur due to shrinkage or swell of expansive soils due to climatic changes, frost expansion of soil, melting of permafrost, slope instability, or other causes. All these factors must be considered during the design of foundations.

Many building codes specify basic foundation design parameters for simple conditions, frequently varying by jurisdiction, but such design techniques are normally limited to certain types of construction and certain types of sites, and are frequently very conservative.

In areas of shallow bedrock, most foundations may bear directly on bedrock; in other areas, the soil may provide sufficient strength for the support of structures. In areas of deeper bedrock with soft overlying soils, deep foundations are used to support structures directly on the bedrock; in areas where bedrock is not economically available, stiff "bearing layers" are used to support deep foundations instead.

Shallow foundations

Shallow foundations are a type of foundation that transfers building load to the layer very near the surface, rather than to a subsurface layer. Shallow foundations typically have a depth to width ratio of less than 1.

Footings

Footings (often called "spread footings" because they spread the load) are structural elements which transfer structure loads to the ground by direct area contact. Footings can be isolated footings for point or column loads, or strip footings for wall or other long (line) loads. Footings are normally constructed from reinforced concrete cast directly onto the soil, and are typically embedded into the ground to penetrate through the zone of frost movement and/or to obtain additional bearing capacity.

Slab foundations

A variant on spread footings is to have the entire structure bear on a single slab of concrete underlying the entire area of the structure. Slabs must be thick enough to provide sufficient rigidity to spread the bearing loads somewhat uniformly, and to minimize differential settlement across the foundation. In some cases, flexure is allowed and the building is constructed to tolerate small movements of the foundation instead. For small structures, like single-family houses, the slab may be less than 300 mm thick; for larger structures, the foundation slab may be several meters thick.

Slab foundations can be either slab-on-grade foundations or embedded foundations, typically in buildings with basements. Slab-on-grade foundations must be designed to allow for potential ground movement due to changing soil conditions, as shown in Fig. 2.1.

Fig. 2.1 Slab-on-grade foundations
图 2.1 分级式板基础

Deep foundations

Deep foundations are used for structures or heavy loads when shallow foundations cannot provide adequate capacity, due to size and structural limitations. They may also be used to transfer building loads past weak or compressible soil layers. While shallow foundations rely solely on the bearing capacity of the soil beneath them, deep foundations can rely on end bearing resistance, frictional resistance along their length, or both in developing the required capacity. Geotechnical engineers use specialized tools, such as the cone penetration test, to estimate the amount of skin and end bearing resistance available in the subsurface.

There are many types of deep foundations including piles, drilled shafts, caissons, piers, and earth stabilized columns. Large buildings such as skyscrapers typically require deep foundations. For example, the Jin Mao Tower in China uses tubular steel piles about 1m (3.3 feet) driven to a depth of 83.5m (274 feet) to support its weight. Fig. 2.2 describes pile driving for a bridge.

In buildings that are constructed and found to undergo settlement, underpinning piles can be used to stabilise the existing building.

Fig. 2.2 Piledriving for a bridge in Napa, California
图 2.2 加利福尼亚纳帕的一座桥在进行沉桩

【Key words】

foundation n. 基础
lateral earth support structure 侧向土支护结构
earth structure 土结构
slope stability 边坡稳定性
marine geotechnical engineering 海洋岩土工程
geosynthetics 土工合成材料
subsurface adj. 表面下的，地下的
consolidation test 固结试验
triaxial shear test 三轴剪切试验
vane shear test 十字板剪切试验
standard penetration test 标准贯入试验
bearing capacity 承载力
settlement n. 沉降
ground movement 地层移动
jurisdiction n. 管辖权；管辖范围；权限；司法权
overlying v. 上覆
shallow foundation 浅基础
spread footing 扩展式基础
reinforced concrete 钢筋混凝土
the zone of frost movement 霜冻运动带
slab foundation 板式基础
slab-on-grade foundation 分级式板基础
embedded foundation 嵌入式基础
deep foundation 深基础
cone penetration test 圆锥贯入试验

tubular *adj.* 管状的
tubular steel pile 钢管桩
pile *n.* 桩
drilled shaft 钻井
caisson *n.* 沉箱基础
pier *n.* 码头，防波堤；桥墩；窗间壁
earth stabilized column 土体加固柱阵
skyscraper *n.* 摩天大楼，超高层大楼；特别高的东西
underpinning pile 托桩（用砖石结构等从下面支撑、加固）

【Translation】

Typical kinds of geotechnical engineering include foundations, lateral earth support structures, earth structures, slope stabilization structures, marine geotechnical engineering structures, geosynthetics products, etc.

典型的岩土工程包括地基基础、挡土墙结构、土工建筑物、边坡加固结构、海洋岩土工程结构、土工合成材料产品等。

Foundations 地基基础

A building's foundation transmits loads from buildings and other structures to the earth. Geotechnical engineers design foundations based on the load characteristics of the structure and the properties of the soils and/or bedrock at the site. In general, geotechnical engineers: ① Estimate the magnitude and location of the loads to be supported; ② Develop an investigation plan to explore the subsurface; ③ Determine necessary soil parameters through field and lab testing (e.g., consolidation test, triaxial shear test, vane shear test, standard penetration test); ④ Design the foundation in the safest and most economical manner.

建筑物的基础把来自建筑物和其他结构的荷载传递到地基上。岩土工程师根据建筑结构的荷载特征以及现场土体和/或基岩的特性来设计基础。在一般情况下，岩土工程师需要：①估计需要承受的荷载的大小和位置；②制定地勘调查计划；③通过现场和实验室测试（例如，固结试验、三轴剪切试验、十字板剪切试验、标准贯入试验）确定必要的土体参数；④按最安全和最经济的原则设计基础。

The primary considerations for foundation support are bearing capacity, settlement, and ground movement beneath the foundations. Bearing capacity is the ability of the site soils to support the loads imposed by buildings or structures. Settlement occurs under all foundations in all soil conditions, though lightly loaded structures or rock sites may experience negligible settlements. For heavier structures or softer sites, **both overall settlement relative to** unbuilt areas or neighboring buildings, **and differential settlement** under a single structure, can be concerns. Of particular concern is settlement which occurs over time, as immediate settlement can usually be compensated for during construction. Ground movement beneath a structure's foundations can occur due to shrinkage or swell of expansive soils due to climatic changes, frost expansion of soil, melting of permafrost, slope instability, or other causes. All these factors must be considered during the design of foundations.

基础支撑能力首先要考虑的是地基承载力、沉降和基础下面的地层移动。地基承载力是现场土体支撑建筑物或构筑物所施加的荷载的能力。无论何种土体条件，任何地基基础都会发生沉降，尽管那些轻微受载的结构或岩石场地可能经受的是可忽略不计的沉降。对于较重的结构或较软弱的现场，既可能涉及相对于未建区域或相邻建筑物的整体沉降，也可能涉及单一建筑物下的不均匀沉降。特别值得关注的是随着时间的推移而产生的沉降，因为瞬时沉降通常会在施工期间获得补偿。由于气候变化、土壤冻胀、冻土融化、边坡失稳或其他原因引起的膨胀土的收缩或膨胀会引起建筑物基础下面的地层移动。在基础设计的过程中必须考虑所有这些因素。

Many building codes specify basic foundation design parameters for simple conditions, frequently varying by jurisdiction, but such design techniques are normally limited to certain types of construction and certain types of sites, and are frequently very conservative.

尽管经常因为管辖权的不同而改变，许多建筑规范还是特别确定了简单情况下基本的基础设计参数，但这种设计技术通常局限于特定类型的建筑和场地，并且常常十分地保守。

In areas of shallow bedrock, most foundations may bear directly on bedrock; **in other areas**, the soil may provide sufficient strength for the support of structures. In areas of deeper bedrock with soft overlying soils, deep foundations are used to support structures directly on the bedrock; in areas where bedrock is not economically available, stiff "bearing layers" are used to support deep foundations instead.

在浅层基岩的地区，大多数基础可能直接建在基岩之上；而在其他一些地区，土体可以提供支撑构筑物的足够强度。在较深基岩上有上覆软土的地区，直接建在基岩上的深基础被用来支撑构筑物；从经济角度出发，没有基岩可用时，就使用硬的"承载层"来支撑深基础。

Shallow foundations 浅基础

Shallow foundations are a type of foundation that transfers building load to the layer very near the surface, rather than to a subsurface layer. Shallow foundations typically have a depth to width ratio of less than 1.

浅基础是一种将建筑物荷载传递到十分接近地表而非地下地层的基础。浅基础深宽比通常小于1。

Footings　"脚"基础

Footings (often called "spread footings" because they spread the load) are structural elements which transfer structure loads to the ground by direct area contact. Footings can be isolated footings for point or column loads, or strip footings for wall or other long (line) loads. Footings are normally constructed from reinforced concrete cast directly onto the soil, and are typically embedded into the ground to penetrate through the zone of frost movement and/or to obtain additional bearing capacity.

"脚"基础（通常称为"扩展式基础"，因为它们能分散压力）属于结构构件，它们通过直接接触面将构筑物荷载传递至地层。"脚"基础可以是承担集中或柱载荷的独立基础，或者是承担墙或其他线性载荷的条形基础。"脚"基础通常由直接浇筑在土体中的钢筋混凝土建成，而且为了获得额外的承载力通常要穿透霜冻运动带与地层结成一体。

Slab foundations　板式基础

A variant on spread footings is to have the entire structure bear on a single slab of concrete underlying the entire area of the structure. Slabs must be thick enough to provide sufficient rigidity to spread the bearing loads somewhat uniformly, and to minimize differential settlement across the foundation. In some cases, flexure is allowed and the building is constructed to tolerate small movements of the foundation instead. For small structures, like single-family houses, the slab may be less than 300 mm thick; for larger structures, the foundation slab may be several meters thick.

扩展式基础的另一种变形是让整个结构由位于整个构筑物以下区域的单块混凝土板来承受。板基础必须足够厚以提供足够的刚度以确保将所承担的荷载以某种方式均匀分散开，并使基础内的不均匀沉降变得最小。在某些情况下，需给予一定的挠度，因而建成的建筑物可以承受基础微小的位移。对于小型建筑物，比如独栋别墅，板基础可能不到300mm厚；但对于较大的构筑物，板基础可能达数米之厚。

Slab foundations can be either slab-on-grade foundations or embedded foundations, typically in buildings with basements. Slab-on-grade foundations must be designed to allow for potential ground movement due to changing soil conditions, as shown in Fig. 2.1.

板基础可以是分级式板基础或嵌入式基础，尤其是对于带地下室的建筑物而言。分级式板基础必须设计成允许承受因为土体条件改变而可能出现的地层移动，如图 2.1 所示。

Deep foundations　深基础

Deep foundations are used for structures or heavy loads when shallow foundations cannot provide adequate capacity, due to size and structural limitations. They may also be used to transfer building loads past weak or compressible soil layers. While shallow foundations rely solely on the bearing capacity of the soil beneath them, deep foundations can rely on end bearing resistance, frictional resistance along their length, or both in developing the required capacity. Geotechnical engineers use specialized tools, such as the cone penetration test, to estimate the amount of skin and end bearing resistance available in the subsurface.

深基础用于那些由于规模和结构限制，浅基础不能为其提供合适的承载力的建筑物，或用于重载荷的情况。它们也可用来实现将建筑物的荷载传递通过软弱或可压缩的土层。浅基础完全依赖于基础下方土体的承载力，而深基础在形成所需要的承载力时能够依靠其端阻力、沿长度方向的摩擦阻力或两者兼而有之。岩土工程师使用专门的工具，比如锥体贯入度试验，来估计地层可提供的表层和端部承阻力。

There are many types of deep foundations including piles, drilled shafts, caissons, piers, and earth stabilized columns. Large buildings such as skyscrapers typically require deep foundations. For example, the Jin Mao Tower in China uses tubular steel piles about 1m (3.3 feet) driven to a depth of 83.5m (274 feet) to support its weight. Fig. 2.2 describes pile driving for a bridge.

深基础有许多类型，包括桩、钻井、沉箱基础、码头和土体加固柱阵。像摩天大楼这样的大型建筑通常需要深基础。例如，中国金茂大厦采用直径约 1m（3.3ft）、嵌入深度达 83.5m（274ft）的钢管桩来支承它的重量。图 2.2 描述了一座桥在进行沉桩。

In buildings that are constructed and found to undergo settlement, underpinning piles can be used to stabilise the existing building.

在建并发现正经历沉降的建筑物，能够使用托桩来加固现有的建筑物。

【Important sentences】

1. Typical kinds of geotechnical engineering include…
   典型的岩土工程包括……（Typical 作定语）
2. both overall settlement relative to…, and differential settlement…
   既可能涉及……的整体沉降，也可能涉及……的不均匀沉降。

2.4 Lateral earth support structures 侧向土支护结构

【Text】

Retaining walls

A retaining wall is a structure that holds back earth. Retaining walls stabilize soil and rock from downslope movement or erosion and provide support for vertical or near-vertical grade changes. Cofferdams and bulkheads, structures to hold back water, are sometimes also considered retaining walls.

The primary geotechnical concern in design and installation of retaining walls is that the retained material is attempting to move forward and downslope due to gravity. This creates soil pressure behind the wall, which can be analysed based on the angle of internal friction (φ) and the cohesive strength (c) of the material and the amount of allowable movement of the wall. This pressure is the smallest at the top and increases toward the bottom in a manner similar to hydraulic pressure, and tends to push the wall forward and overturn it. Groundwater behind the wall that is not dissipated by a drainage system causes an additional horizontal hydraulic pressure on the wall.

Gravity walls

Gravity walls depend on the size and weight of the wall mass to resist pressures from behind. Gravity walls will often have a slight setback, or batter, to improve wall stability. For short, landscaping walls, gravity walls made from dry-stacked (mortarless) stone or segmental concrete units (masonry units) are commonly used.

Earlier in the 20th century, taller retaining walls were often gravity walls made from large masses of concrete or stone. Today, taller retaining walls are increasingly built as composite gravity walls such as: geosynthetic or steel-reinforced backfill soil with precast facing, gabions (stacked steel wire baskets filled with rocks), crib walls (cells built up log cabin style from precast concrete or timber and filled with soil or free draining gravel) or soil-nailed walls (soil reinforced in place with steel and concrete rods), shown in Fig. 2.3.

For reinforced-soil gravity walls, the soil reinforcement is placed in horizontal layers throughout the height of the wall. Commonly, the soil reinforcement is geogrid, a high-strength polymer mesh, which provide tensile strength to hold soil together. The wall face is often of precast, segmental concrete units that can tolerate some differential movement. The reinforced soil's mass, along with the facing, becomes the gravity wall. The reinforced mass must be built large enough to retain

Fig. 2.3 Gabions and grib walls
图 2.3 石笼和格笼墙

the pressures from the soil behind it. Gravity walls usually must be a minimum of 30 to 40 percent as deep (thick) as the height of the wall, and may have to be larger if there is a slope or surcharge on the wall.

Cantilever walls

Prior to the introduction of modern reinforced-soil gravity walls, cantilever walls were the most common type of taller retaining wall, as shown in Fig. 2.4. Cantilever walls are made from a relatively thin stem of steel-reinforced, cast-in-place concrete or mortared masonry (often in the shape of an inverted T). These walls cantilever loads (like a beam) to a large, structural footing; converting horizontal pressures from behind the wall to vertical pressures on the ground below. Sometimes cantilever walls are buttressed on the front, or include a counterfort on the back, to improve their stability against high loads. Buttresses are

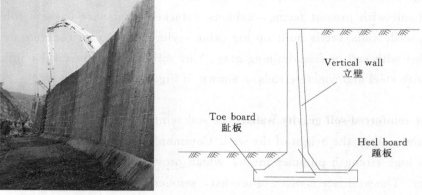

Fig. 2.4 Cantilever walls
图 2.4 悬臂式墙

short wing walls at right angles to the main trend of the wall. These walls require rigid concrete footings below seasonal frost depth. This type of wall uses much less material than a traditional gravity wall.

Cantilever walls resist lateral pressures by friction at the base of the wall and/or passive earth pressure, the tendency of the soil to resist lateral movement.

Basements are a form of cantilever walls, but the forces on the basement walls are greater than on conventional walls because the basement wall is not free to move.

Excavation shoring

Shoring of temporary excavations frequently requires a wall design which does not extend laterally beyond the wall, so shoring extends below the planned base of the excavation. Common methods of shoring are the use of sheet piles or soldier beams and lagging. Sheet piles are a form of driven piling using thin interlocking sheets of steel to obtain a continuous barrier in the ground, and are driven prior to excavation. Soldier beams are constructed of wide flange steel H sections spaced about 2 – 3 m apart, driven prior to excavation. As the excavation proceeds, horizontal timber or steel sheeting (lagging) is inserted behind the H pile flanges.

In some cases, the lateral support which can be provided by the shoring wall alone is insufficient to resist the planned lateral loads; in this case additional support is provided by walers or tie-backs. Walers are structural elements which connect across the excavation so that the loads from the soil on either side of the excavation are used to resist each other, or which transfer horizontal loads from the shoring wall to the base of the excavation. Tie-backs are steel tendons drilled into the face of the wall which extend beyond the soil which is applying pressure to the wall, to provide additional lateral resistance to the wall. Sheet piles and lagging is shown in Fig. 2.5.

【Key words】

retaining wall 挡土墙
cofferdam $n.$ 围堰
bulkhead $n.$ 防水墙（壁）
soil pressure 土压力
angle of internal friction (φ) 内摩擦角 (φ)
cohesive strength (c) 黏聚力 (c)
hydraulic pressure 水压力
overturn $vt.$ & $vi.$ （使）翻倒，倾覆 $n.$ 推翻，垮台；瓦解；灭亡，毁灭
groundwater $n.$ 地下水
dissipate $v.$ 消散

(a) Sheet piles 　　　　　　　　(b) Lagging
　　钢板桩　　　　　　　　　　　推拉杆

Fig. 2.5　Sheet piles and lagging
图 2.5　钢板桩和推拉杆

gravity wall　重力墙
setback　n. 缩进
dry-stacked (mortarless) stone　石块垒砌（不用灰浆）
masonry unit　砌筑块材
precast facing　预制面
gabion　n. 石笼
crib wall　格笼墙
soil-nail wall　土钉墙
geogrid　n. 土工格栅
cantilever wall　悬臂式墙
buttress　n. 扶壁；支撑物　vt. 支持，鼓励；用扶壁支撑，加固
counterfort　n. 护墙

【Translation】

Retaining walls　挡土墙

A retaining wall is a structure that holds back earth. Retaining walls stabilize soil and rock from downslope movement or erosion and provide support for vertical or near-vertical grade changes. Cofferdams and bulkheads, structures to hold back water, are sometimes also considered retaining walls.

挡土墙是一种能够阻挡后方土体的结构。挡土墙能阻止岩土向坡下滑动或受侵蚀，并为垂直或近似垂直的梯度改变提供支撑。围堰、阻流坝和用来挡水的水工建筑物，有时也可归为挡土墙。

The primary geotechnical concern in design and installation of retaining walls is that the retained material is attempting to move forward and downslope due to gravity. This creates soil pressure behind the wall, which can be analysed based on the angle of internal friction (φ) and the cohesive strength (c) of the material and the amount of allowable movement of the wall. This pressure is the **smallest** at the top and increases toward the bottom **in a manner similar to** hydraulic pressure, and **tends to** push the wall forward and overturn it. Groundwater **behind the wall that is not dissipated by** a drainage system causes an additional horizontal hydraulic pressure on the wall.

在挡土墙的设计及建造过程中，岩土工程首要关注的是挡土墙材料由于重力作用试图向前和向下滑动的趋势。这在挡土墙背后产生土压力，可以根据材料的内摩擦角（φ）、黏聚力（c）以及墙体允许的位移量来分析。这种压力在顶部值最小，以一种类似于水压力的方式向下递增，同时有向前推动挡土墙并推翻它的倾向。挡土墙后面未能由排水系统疏干的地下水会对墙壁产生一个附加的水平水压力。

Gravity walls 重力墙

Gravity walls depend on the size and weight of the wall mass to resist pressures from behind. Gravity walls will often have a slight setback, or batter, to improve wall stability. **For short**, landscaping walls, gravity walls made from dry-stacked (mortarless) stone or segmental concrete units (masonry units) are commonly used.

重力墙依靠墙体的大小和重量抵抗来自墙背后的压力。重力墙往往会通过向后轻微地缩进或内倾来提高墙体稳定性。为了简便，由石块垒砌（不用灰浆）或混凝土砌块建成的景观墙、重力墙也经常使用。

Earlier in the 20th century, taller retaining walls were often gravity walls made from large masses of concrete or stone. Today, taller retaining walls are increasingly built as composite gravity walls such as: geosynthetic or steel-reinforced backfill soil with precast facing, gabions (stacked steel wire baskets filled with rocks), crib walls (cells built up log cabin style from precast concrete or timber and filled with soil or free draining gravel) or soil-nailed walls (soil reinforced in place with steel and concrete rods), shown in Fig. 2.3.

在 20 世纪初期，较高的挡土墙通常是用大量的混凝土或石头建成的重力墙。现今，较高的挡土墙越来越多地被建成为复合重力墙，比如：土工合成材料或通过预制面钢筋加固回填土；石笼（由装满岩石的钢丝笼堆叠）；格笼墙（用预制混凝土或木料搭建木屋式的单元并用土或排水砾石填充）或土钉墙（用钢筋混凝土杆现场加固土体），如图 2.3 所示。

For reinforced-soil gravity walls, the soil reinforcement is placed in horizontal layers throughout the height of the wall. Commonly, the soil reinforcement is geogrid, a high-strength polymer mesh, that provide tensile strength to hold soil together. The wall face is often of precast, segmental concrete units that can tolerate some differential movement. The reinforced soil's mass, along with the facing, becomes the gravity wall. The reinforced mass must be built large enough to retain the pressures from the soil behind it. Gravity walls usually must be a minimum of 30 to 40 percent as deep (thick) as the height of the wall, and may have to be larger if there is a slope or surcharge on the wall.

对于加固土的重力墙，沿着整个墙体的高度，水平分层式地放置土体加固装置。通常，土体加固装置是一种土工格栅，一种高强度的聚合物网格，可以提供抗拉强度将土体捆住。墙面通常是预制的、分段式的混凝土单元体，它可以允许有一些不均匀位移。沿着覆盖面的加固土体成为重力墙。加固体必须建得足够大以抵抗来自墙体背后土体的压力。重力墙的最小深度（厚度）通常必须是墙体高度的30%～40%，如果面对的是一个斜坡或者墙体上有附加载荷，则可能不得不取更大值。

Cantilever walls 悬臂式墙

Prior to the introduction of modern reinforced-soil gravity walls, cantilever walls were the most common type of taller retaining wall, as shown in Fig. 2.4. Cantilever walls are made from a relatively thin stem of steel-reinforced, cast-in-place concrete or mortared masonry (often in the shape of an inverted T). These walls cantilever loads (like a beam) to a large, structural footing; converting horizontal pressures from behind the wall to vertical pressures on the ground below. Sometimes cantilever walls are buttressed on the front, or include a counterfort on the back, to improve their stability against high loads. Buttresses are short wing walls at right angles to the main trend of the wall. These walls require rigid concrete footings below seasonal frost depth. This type of wall uses much less material than a traditional gravity wall.

在引入现代加固土重力墙之前，悬臂式墙是较高挡土墙中最常见的类型，如图2.4所示。悬臂墙是由相对较细的加固钢筋做骨架，通过现浇混凝土或浆砌而成（通常呈倒T形状）。这些墙通过悬臂（就像一根梁）将载荷传递至大的结构基础上；将来自墙后的水平压力转换成作用于地下的垂直压力。有时候悬臂式墙需在前端用扶壁支撑加固，或在墙的背面设置护墙，以便提高它们抵抗高负荷时的稳定性。外支墙都是一些短翼墙，与主墙的走向垂直相交。这些墙需要在季节性的冻结深度之下设置混凝土刚性基础。与传统的重力墙相比较，这种类型的墙需用的材料要少得多。

Cantilever walls resist lateral pressures by friction at the base of the wall and/or passive earth pressure, the tendency of the soil to resist lateral movement.

悬臂式墙通过墙基底部的摩擦力和/或被动土压力来抵抗侧向压力，被动土压力是指土体抵抗侧向运动的趋势。

Basements are a form of cantilever walls, but the forces on the basement walls are greater than on conventional walls because the basement wall is not free to move.

地下室是悬臂墙的一种形式，但作用于地下室墙上的力都大于作用在常规墙体上的力，因为地下室墙是不能自由移动的。

Excavation shoring 基坑支护

Shoring of temporary excavations frequently requires a wall design which does not extend laterally beyond the wall, so shoring extends below the planned base of the excavation. Common methods of shoring are the use of sheet piles or soldier beams and lagging. Sheet piles are a form of driven piling using thin interlocking sheets of steel to obtain a continuous barrier in the ground, and are driven prior to excavation. Soldier beams are constructed of wide flange steel H sections spaced about 2–3 m apart, driven prior to excavation. As the excavation proceeds, horizontal timber or steel sheeting (lagging) is inserted behind the H pile flanges.

临时开挖支护通常要求设计的挡墙不能向侧向扩展，因此支护墙体只能向规划的开挖基底以下延伸。常用的支护方法有板桩或立柱以及推拉杆。板桩是一种贯入式桩，为获得一种连续屏障，将薄的、连锁的钢片贯入地下，并且是在开挖之前贯入。基坑围护立柱由宽翼"H"型钢构成，其间距约为2～3m，在开挖前贯入。随着开挖工作的进行，水平的木支护或钢板桩（推杆）被嵌入到宽翼"H"形钢桩的背后。

In some cases, the lateral support which can be provided by the shoring wall alone is insufficient to resist the planned lateral loads; in this case additional support is provided by walers or tie-backs. Walers are structural elements which connect across the excavation so that the loads from the soil on either side of the excavation are used to resist each other, or which transfer horizontal loads from the shoring wall to the base of the excavation. Tie-backs are steel tendons drilled into the face of the wall which extend beyond the soil which is applying pressure to the wall, to provide additional lateral resistance to the wall. Sheet piles and lagging is shown in Fig. 2.5.

在某些情况下，由支护墙体单独提供的侧向支护力不足以抵抗设计的侧向荷载；在这种情况下，额外的支撑力就由横撑或回拉杆提供。横撑是结构部件，它横跨开挖空间实现连接，因此，来自开挖空间两边的土体荷载被用来相互支撑，或者可以将来自挡墙的水平荷载传递至开挖空间底部。回拉杆是钻入延伸至土体背面的墙面的钢索，用来对墙体施加压力，从而对墙体提供额外的侧向抵抗力。钢板桩和推拉杆如图 2.5 所示。

【Important sentences】

1. This pressure is smallest at the top and increases toward the bottom in a manner similar to hydraulic pressure, and tends to push the wall forward and overturn it.

   smallest 表示最高级；in a manner similar to… 以一种类似于……的方式；tends to do sth. 表示"趋向于"。

2. Groundwater behind the wall that is not dissipated by a drainage system causes an additional horizontal hydraulic pressure on the wall.

   that 引导定语从句修饰 Groundwater，behind the wall 为 Groundwater 的位置状语，表示"墙后的地下水"；is not dissipated by 表被动。句子的主干部分为：Groundwater causes an additional horizontal hydraulic pressure 地下水产生一个附加的水平水压力。

3. For short

   为简便起见，为了简便；相当于 in short。

2.5 Earth structures 土工建筑物

【Text】

Earthworks (engineering)

Earthworks are engineering works created through the moving or processing of parts of the earth's surface involving quantities of soil or unformed rock. The earth may be moved to another location and formed into a desired shape for a purpose. Much of earthworks involves machine excavation and fill or backfill. Fig. 2.6 is a used Caterpillar bulldozer.

Fig. 2.6 A used Caterpillar bulldozer
图 2.6 二手的 Caterpillar 推土机

Types of excavation

Excavation may be classified by type of material:

- Topsoil excavation
- Earth excavation

2.5 Earth structures 土工建筑物

- Rock excavation
- Muck excavation—this usually contains excess water and unsuitable soil
- Unclassified excavation—this is any combination of material types

Excavation may be classified by the purpose:

- Stripping
- Roadway excavation
- Drainage or structure excavation
- Bridge excavation
- Channel excavation
- Footing excavation
- Borrow excavation
- Dredge excavation

Civil engineering use

Typical earthworks include roads, railway beds, causeways, dams, levees, canals, and berms. Other common earthworks are land grading to reconfigure the topography of a site, or to stabilize slopes.

Military use

In military engineering, earthworks are, more specifically, types of fortifications constructed from soil. Although soil is not very strong, it is cheap enough that huge quantities can be used, generating formidable structures. Examples of older earthwork fortifications include moats, sod walls, motte-and-bailey castles, and hill forts. Modern examples include trenches and berms.

Equipment

Heavy construction equipment is usually used due to the amounts of material to be moved—up to millions of cubic metres. Earthwork construction was revolutionised by the development of the (Fresno) scraper and other earth-moving machines such as the loader, production trucks, the grader, the bulldozer, the backhoe, and the dragline excavator.

Mass haul planning

Engineers need to concern themselves with issues of geotechnical engineering (such as soil density and strength) and with quantity estimation to ensure that soil volumes in the cuts match those of the fills, while minimizing the distance of movement. In the past, these calculations were done by hand using a slide rule and with methods such as Simpson's rule.

Now they can be performed with a computer and specialized software, including optimisation on haul cost and not haul distance (as haul cost is not proportional to haul distance).

Slope stability

Slope stability is the analysis of soil covered slopes and its potential to undergo movement. Stability is determined by the balance of shear stress and shear strength. A previously stable slope may be initially affected by preparatory factors, making the slope conditionally unstable. Triggering factors of a slope failure can be climatic events because they can make a slope actively unstable, leading to mass movements. Mass movements can be caused by increases in shear stress, such as loading, lateral pressure, and transient forces. Alternatively, shear strength may be decreased by weathering, changes in pore water pressure, and organic material. Simple slope slip model is shown in Fig. 2.7.

Fig. 2.7　Simple slope slip model
图 2.7　简单边坡滑动模型

Marine geotechnical engineering

In subsea geotechnical engineering, seabed materials are considered a two-phase material composed of ① rock or mineral particles and ② water. Structures may be fixed in place in the seabed—as in piers, jetties, or fixed-bottom wind turbines—or may be floating structures anchored to remain in a sea-surface position that remain roughly fixed relative to its geotechnical anchor point.

Undersea foundations

Examples of undersea foundations include multiple-pile foundations as used in many piers and monopile foundations used for many fixed-bottom offshore wind turbines.

Floating-moored structures

A tension leg mooring system for a wind turbine: left-hand tower-bearing structure (grey) is free floating, the right-hand structure is pulled by the tensioned cables (red) down towards the seabed anchors (light-grey), as shown in Fig. 2.8.

Undersea mooring of human-engineered floating structures include a large number of offshore oil and gas platforms and, since 2008, a few floating wind turbines.

Two common types of engineered design for anchoring floating structures include tension-leg and catenary loose mooring systems. "Tension leg mooring systems have vertical tethers under tension providing large restoring moments in pitch and roll. Catenary mooring systems provide station keeping for an offshore structure yet provide little stiffness at low tensions." A floating wind turbine moored is shown in Fig. 2.9.

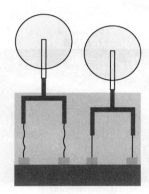
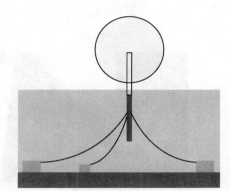

Fig. 2.8　A tension leg mooring system for a wind turbine

图 2.8　用于风力发电机的张力腿系泊系统

Fig. 2.9　A floating wind turbine moored by catenary cables

图 2.9　悬链浮式风力发电机系泊系统

A third form of mooring system is the ballasted catenary configuration, created by adding multiple-tonne weights hanging from the midsection of each anchor cable in order to provide additional cable tension and therefore increase stiffness of the above-water floating structure.

Geosynthetics

Geosynthetics is the umbrella term used to describe a range of synthetic products used to aid in solving some geotechnical problems. The term is generally regarded to encompass four main products: geotextiles, geogrids, geomembranes, and geocomposites (Fig. 2.10). The synthetic nature of the products make them suitable for use in the ground where high levels of durability are required, though this is not to say that they are indestructible. Geosynthetics are available in a wide range of forms and materials, each to suit a slightly different end use. These products have a wide range of applications and are currently used in many civil and geotechnical engineering applications including roads, airfields, railroads, embankments, retaining structures, reservoirs, canals, dams, landfills, bank protection and coastal engineering.

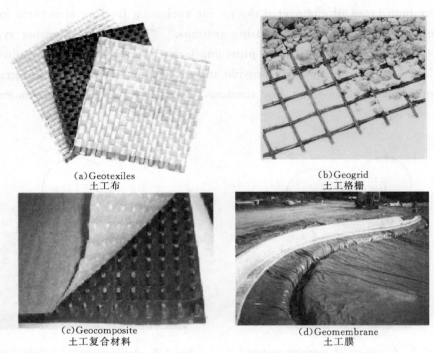

Fig. 2.10　Geosynthetic products
图 2.10　土工合成材料产品

【Key words】

excavation *n.* 挖掘，发掘
backfill *vt.* 回填；装填
muck *n.* 淤泥；垃圾；肥料；品质低劣的东西
stripping *n.* 剥离；剥脱；拆封
causeway *n.* 堤道；铺道
topography *n.* 地势；地形学；地志
fortification *n.* 设防；防御工事；加强
motte-and-bailey　一种木制城堡
hill fort　山丘堡垒
trench *n.* 沟，沟渠；战壕；堑壕
berm *n.* 护坡道
revolutionise *vt.* 彻底改变（等于 revolutionize）；使革命化
Simpson's rule　辛普森积分法；辛普森法则
slope failure　边坡破坏；滑坡；山泥倾泻；斜坡崩塌
pore water pressure　孔隙水压力
pier *n.* 码头，直码头；桥墩；窗间壁
jetty *n.* 码头；防波堤
wind turbine　风力涡轮机；风力发电机

mooring system　系泊系统；锚泊系统
synthetic *adj*. 综合的；合成的，人造的
reservoir *n*. 水库；蓄水池

【Translation】

Earthworks（engineering）　土方（工程）

Earthworks are engineering works created through the moving or processing of parts of the earth's surface involving quantities of soil or unformed rock. The earth may be moved to another location and formed into a desired shape for a purpose. Much of earthworks involves machine excavation and fill or backfill. Fig. 2.6 is a used Caterpillar bulldozer.

土方工程是指通过搬运或处置地表土石而产生的工程作业，涉及数量巨大的土体或未成形的岩石。土体可以被搬运到其他位置，并根据需要堆成想要的样子。许多土方工程都涉及机械开挖、填充或回填。图 2.6 为二手 Caterpillar 推土机照片。

Types of excavation　开挖类型

Excavation may be classified by type of material：

开挖可根据开挖材料类型进行分类：

- Topsoil excavation
  表层土开挖
- Earth excavation
  土体开挖
- Rock excavation
  岩石开挖
- Muck excavation—this usually contains excess water and unsuitable soil
  淤泥开挖——通常包含大量的水和不好用的土
- Unclassified excavation—this is any combination of material types
  未归类开挖——指各类材料混合在一起的情况

Excavation may be classified by the purpose：

开挖可根据目的来分类：

- Stripping
  剥离
- Roadway excavation

- Drainage or structure excavation
  排水或结构开挖
- Bridge excavation
  桥梁开挖
- Channel excavation
  通道开挖
- Footing excavation
  基础开挖
- Borrow excavation
  取土开挖
- Dredge excavation
  疏浚开挖

Civil engineering use 土木工程应用

Typical earthworks include roads, railway beds, causeways, dams, levees, canals, and berms. Other common earthworks are **land grading** to **reconfigure** the topography of a site, or to **stabilize slopes.**

典型的土方工程包括公路、铁路路基、堤道、水坝、防洪堤、运河和护坡道。其他常见的土方工程包括为改变现场地貌，或为了边坡稳定而进行的土地阶梯化工作。

Military use 军事用途

In military engineering, earthworks are, more specifically, types of fortifications constructed from soil. **Although** soil is not very strong, **it is cheap enough that** huge quantities can be used, **generating** formidable structures. Examples of older earthwork fortifications include moats, sod walls, motte-and-bailey castles, and hill forts. Modern examples include trenches and berms.

在军事工程中，土方工程更具体地说，指的是各类由土体构建的防御工事。虽然土体不是很坚硬，但它足够廉价因而可大量使用，并形成强大的结构。较早期的土方防御工程案例包括护城河、草皮墙、有灌木丛和外墙的城堡以及山堡。现代的案例则包括战壕和护坡道。

Equipment 设施装备

Heavy construction equipment is usually used due to the amounts of material to be moved—up to millions of cubic metres. Earthwork construction was revolutionised by the development of the scraper and other earth-moving machines such as the loader, production trucks, the grader, the bulldozer, the backhoe, and the dragline excavator.

因为被搬运的材料量多到数百万立方米，通常要使用重型建筑设备。土方施工由于铲土机和其他运土机械，比如装载机、载重卡车、平地机、推土机、反铲挖土机和拉索挖掘机等的发展而发生了变革。

Mass haul planning　　岩土搬运优化

Engineers need to concern themselves with issues of geotechnical engineering (such as soil density and strength) and with quantity estimation to ensure that soil volumes in the cuts match those of the fills, while minimizing the distance of movement. In the past, these calculations were done by hand using a slide rule and with methods such as Simpson's rule.

工程师们需要自觉关注的问题既有土壤密度和强度等岩土工程问题，还有在最大限度地减少搬运距离的同时，为保证挖掘的土量与填充需要的土量相匹配的土量估计问题。在过去，这些计算工作都是使用计算尺手算完成的，以及借助辛普森法则这类算法来完成。

Now they can be performed with a computer and specialized software, including optimization on haul cost and not haul distance (as haul cost is not proportional to haul distance).

现在，这些计算工作可以使用计算机和专业软件来完成，包括根据搬运成本但不计搬运距离（因为搬运成本与搬运距离并不成正比）进行优化。

Slope stability　　边坡稳定性

Slope stability is the analysis of soil covered slopes and its potential to undergo movement. Stability is determined by the balance of shear stress and shear strength. A previously stable slope may be initially affected by preparatory factors, making the slope conditionally unstable. Triggering factors of a slope failure can be climatic events because they can make a slope actively unstable, leading to mass movements. Mass movements can be caused by increases in shear stress, such as loading, lateral pressure, and transient forces. Alternatively, shear strength may be decreased by weathering, changes in pore water pressure, and organic material. Simple slope slip section is shown in Fig. 2.7.

边坡稳定性研究包括对覆土边坡及其可能经受的滑动的分析。边坡稳定性由剪应力和抗剪强度的平衡来决定。一个原本稳定的斜坡可能会受到某些预设因素的影响，造成边坡出现条件性的不稳定。边坡失稳的触发因素可以是气候事件，因为它能促使边坡变得活跃而不稳定，进而导致坡体滑动。坡体滑动可能是由于剪切应力的增加所致，比如加载、侧向压力和瞬时力。另外，剪切强度可能会因为风化作用、孔隙水压力的变化和有机材料等影响而降低。简单边坡滑动剖面如

图 2.7 所示。

Marine geotechnical engineering　海洋岩土工程

In subsea geotechnical engineering, seabed materials are considered a two-phase material composed of rock or mineral particles and water. Structures may be fixed in place in the seabed—as in piers, jetties, or fixed-bottom wind turbines—or may be floating structures anchored to remain in a sea-surface position that remain roughly fixed relative to its geotechnical anchor point.

在海下岩土工程中，海底材料被认为是一种两相材料，由岩石或矿物颗粒和水组成。构筑物可以固定在海底某个位置，就像码头、防波堤或底部固定的风轮机那样，或者是可漂浮构筑物靠锚固而保持在海面某个位置，停留的位置相对于岩土锚固点而言是大致固定的。

Undersea foundations　海底基础

Examples of undersea foundations include multiple-pile foundations as used in many piers and monopile foundations used for many fixed-bottom offshore wind turbines.

海下基础包括在很多码头使用的复合桩基础，以及用于许多底部固定的离岸风力发动机的单桩基础。

Floating-moored structures　浮式系泊结构

A tension leg mooring system for a wind turbine: left-hand tower-bearing structure is free floating, the right-hand structure is pulled by the tensioned cables down towards the seabed anchors, as shown in Fig. 2.8.

用于风力发电机的张力腿系泊系统：左边的承载塔结构是自由浮动的，右边的结构则由拉索向着海底锚头方向拉紧，如图 2.8 所示。

Undersea mooring of human-engineered floating structures include a large number of offshore oil and gas platforms and, since 2008, a few floating wind turbines. Two common types of engineered design for anchoring floating structures include tension-leg and catenary loose mooring systems. "Tension leg mooring systems have vertical tethers under tension providing large restoring moments in pitch and roll. Catenary mooring systems provide station keeping for an offshore structure yet provide little stiffness at low tensions." A floating wind turbine moored is shown in Fig. 2.9.

海底系泊人类工程浮体结构包括大量离岸油、气平台，以及自 2008 年以来的几座浮式风力发电机组。用来锚固浮式构筑物的两种常用工程设计包括张力腿式和悬链松弛式系泊系统。在海水涌动条件下，张力腿系泊系统拥有在张力下能提供巨大复原力矩的垂直的绳索。悬链系泊系统为保护离岸构筑物提供了平台，在低张力条件下也提供较小的刚度。悬链浮式风力发电机系泊系统如图 2.9 所示。

A third form of mooring system is the ballasted catenary configuration, created by adding multiple-tonne weights hanging from the midsection of each anchor cable in order to provide additional cable tension and therefore increase stiffness of the above-water floating structure.

第三种形式的系泊系统是压舱式悬链结构，为了提供额外的缆索拉力，通过在每根锚索中段悬挂数吨重的重物来实现，也因此提高了水上浮式构筑物的刚度。

Geosynthetics　土工合成材料

Geosynthetics is the umbrella term used to describe a range of synthethic products used to aid in solving some geotechnical problems. The term is generally regarded to encompass four main products: geotextiles, geogrids, geomembranes, and geocomposites (Fig. 2.10). The synthetic nature of the products make them suitable for use in the ground where high levels of durability are required, though this is not to say that they are indestructible. Geosynthetics are available in a wide range of forms and materials, each to suit a slightly different end use. These products have a wide range of applications and are currently used in many civil and geotechnical engineering applications including roads, airfields, railroads, embankments, retaining structures, reservoirs, canals, dams, landfills, bank protection and coastal engineering.

土工合成材料是用来描述一系列合成产品的总称，被用来帮助解决一些岩土工程问题。该术语一般认为包含四种主要产品：土工织物，土工格栅，土工膜和土工复合材料（图 2.10）。这些产品的合成性质使它们适用于对耐用性有较高要求的地面，但这并不是说它们是坚不可摧的。土工合成材料的形式和材料可选择范围大，每种的用途都稍有不同。这些产品有着广泛的应用，目前被用在许多土木工程和岩土工程中，包括公路、机场、铁路、路堤岸堤、支护结构、水库、运河、堤坝、垃圾填埋场、护岸及海岸工程。

【Important sentences】

1. Excavation may be classified by…
   开挖可根据……进行分类。
2. Other common earthworks are land grading to reconfigure the topography of a site, or to stabilize slopes.
   land grading　土方工程中的土地阶梯化工作
   configure　安装、使成形

reconfigure 重新配置
stabilize slopes 稳定边坡

3. Although soil is not very strong, it is cheap enough that huge quantities can be used, generating formidable structures.

Although 表转折，句子的主干部分为 it is cheap enough that…。generating 引导定语从句，修饰 can be used。

Chapter 3
Mechanical Properties of Soil/Rock and Testing

岩土力学特性与测试

3.1 Definition of soil 土体定义

【Text】

Soil is the mixture of minerals, organic matter, gases, liquids, and the myriad of organisms that together support plant life. It is a natural body that exists as part of the pedosphere and which performs four important functions: it is a medium for plant growth; it is a means of water storage, supply and purification; it is a modifier of the atmosphere; and it is a habitat for organisms that take part in decomposition of organic matter and the creation of a habitat for new organisms.

Soil is considered to be the "skin of the earth" with interfaces between the lithosphere, hydrosphere, atmosphere, and biosphere. Soil consists of a solid phase (minerals and organic matter) as well as a porous phase that holds gases and water. Accordingly, soils are often treated as a three-state system.

Soil is the end product of the influence of the climate, relief (elevation, orientation, and slope of terrain), biotic activities (organisms), and parent materials (original minerals) interacting over time. Soil continually undergoes development by way of numerous physical, chemical and biological processes, which include weathering with associated erosion. Most soils have a density between 1 g/cm^3 and 2 g/cm^3.

Soil science has two main branches of study: Edaphology and Pedology. Pedology is focused on the formation, description (morphology), and classification of soils in their natural environment, whereas Edaphology is concerned with the influence of soils on organisms. In engineering terms, soil is referred to as regolith, or loose rock material that lies above the "solid geology". Soil is commonly referred to as "earth" or "dirt"; technically, the term "dirt" should be restricted to displaced soil. Fig. 3.1 shows the distribution of soil and rock layer.

As a perspective of the classification of soil, the soil type can be divided into organic matter soil, peat soil, silt soil, cohesive soil, powder soil, sand and gravel soil, etc. The category of soil is generally determined by the following indicators: the soil particle composition and characteristics, natural water content of soil, soil plasticity index (liquid limit, plastic limit and plastic index) and organic matter in soil, etc.

As soil resources serve as a basis for food security, the international community advocates for its sustainable and responsible use through different types of soil governance.

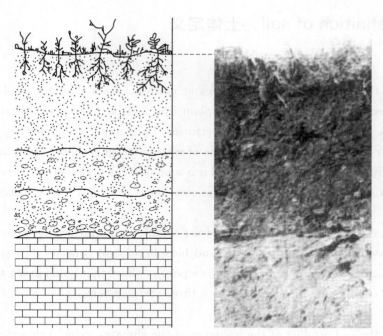

Fig. 3.1　Distribution of soil and rock layer
图 3.1　岩层土层分布图
A—soil; B—laterite, a regolith; C—saprolite, a less-weathered regolith
A—土; B—红土带, 一种风化层; C—腐泥岩, 风化较少的风化层

【Key words】

myriad *adj.* 无数的；多种的，各式各样的
organism *n.* 有机体；生物体
pedosphere *n.* （地球的）土壤圈
purification *n.* 净化；提纯；纯化
modifier *n.* 修改器，调节器
atmosphere *n.* 大气，空气；大气层
decomposition *n.* 分解；腐烂
interface *n.* 界面；交界面
lithosphere *n.* 岩石圈，陆界
hydrosphere *n.* 水圈，水气，水界
biosphere *n.* 生物圈；生物界；生物层
solid phase *n.* 固相，固态
end product　最终结果，最终产物
relief *n.* 地形
elevation *n.* 海拔；标高；高度
orientation *n.* 方向；定向；定位
slope of terrain　地形斜坡面
biotic *adj.* 生命的；生物的

parent materials 母质，原生材料
weathering n. 侵蚀，风化；雨蚀
erosion n. 腐蚀，侵蚀，磨损；烧蚀
Pleistocene adj. 更新世的 n. 更新世
Cenozoic adj. 新生界的 n. 新生世
fossilize v. 使成化石
archean n. 太古宙 adj. 太古代的
edaphology n. 土壤生态学；农业土壤学
pedology n. 土壤学；发生土壤学
morphology n. 形态学
regolith n. 风化层，土被；浮土

【Translation】

Soil is the mixture of minerals, organic matter, gases, liquids, and the myriad of organisms that together support plant life. **It is** a natural body **that** exists as part of the pedosphere and **which** performs four important functions: **it is** a medium for plant growth; **it is** a means of water storage, supply and purification; **it is** a **modifier** of the atmosphere; **and it is** a habitat for organisms **that take part in** decomposition of organic matter **and** the creation of a habitat for new organisms.

土是矿物、有机质、气体、液体和无数生物的集合体，它们共同承担植物生长。土是作为土壤圈的一部分而存在的自然体，它具有四种重要的作用：它是植物生长的培养基；它是储存、供给和净化水的一种途径；它是大气的调节器；它也是那些参与有机质分解的生物的栖息地，同时是新生物的诞生场所。

Soil is considered to be the "skin of the earth" with interfaces between the lithosphere, hydrosphere, atmosphere, and biosphere. Soil consists of a solid phase (minerals and organic matter) **as well as** a porous phase **that holds gases and water**. Accordingly, soils **are often treated as a three-state system**.

土被认为是"地球的皮肤"，它是介于岩石圈、水圈、大气圈和生物圈之间的界面。土是由固相物质（矿物和有机质）和能包含气体和水的多孔相物质组成。相应地，土通常被作为三相体系对待。

Soil is the end product of the influence of the climate, relief (elevation, orientation, and slope of terrain), biotic activities (organisms), and parent materials (original minerals) interacting over time. Soil continually undergoes development by way of numerous physical, chemical and biological processes, which include weathering with associated erosion. Most soils have a density between 1 and 2 g/cm^3.

土是由气候、地形地貌（当地的海拔、方向和坡度）、生物活动（生物）和原生材料（原生矿物）等随着时间产生相互作用的影响下形成的最终产物。土持续经受大量物理、化学和生物变化层面的进化，这包括伴侵蚀性风化作用。大多数土的密度介于 $1\sim 2g/cm^3$。

Soil science has two main branches of study: Edaphology and Pedology. Pedology is focused on the formation, description (morphology), and classification of soils in their natural environment, whereas Edaphology is concerned with the influence of soils on organisms. In engineering terms, soil **is referred to as** regolith, or loose rock material that lies above the "solid geology". Soil is commonly referred to as "earth" or "dirt"; technically, the term "dirt" should be restricted to displace soil. Fig. 3.1 shows the distribution of soil and rock layer.

土壤科学有两个主要的研究分支：土壤生态学和土壤学。土壤学专注于自然环境中土的形成、描述（形态）和分类，而土壤生态学则是注重土对生物的影响。在工程学上，土是指风化层，或者那些"坚硬地质体"上覆的松散岩石材料。土通常被认为是"土壤"或者"污泥"，但严格来说，"污泥"这个词应该禁止用来替代"土"这个词。图3.1为岩层土层分布图。

As a perspective of the classification of soil, the soil type can be divided into organic matter soil, peat soil, silt soil, cohesive soil, powder soil, sand and gravel soils, etc. The category of soil is generally determined by the following indicators: the soil particle composition and characteristics, natural water content of soil, soil plasticity index (liquid limit, plastic limit and plastic index) and organic matter in soil, etc.

从土的分类来看，土可分为有机质土、泥炭、淤泥土、黏性土、粉土、砂土和碎石土等。土的类别一般由下列指标来确定：土的颗粒组成及其特征，土的天然含水量，土的塑性指标（液限、塑限和塑性指数）以及土中有机物存在情况等。

As soil resources serve **as** a basis for food security, the international community advocates for its sustainable and responsible use through different types of soil governance.

土地资源是食物安全的基础，国际社会提倡通过不同的土地管理方法来实现土地使用的可持续性和责任性。

【Important sentences】

1. It is a natural body that exists as part of the pedosphere and which performs four important functions: it is a medium for plant growth; it is a means of water

storage, supply and purification; it is a modifier of the atmosphere; and it is a habitat for organisms that take part in decomposition of organic matter and the creation of a habitat for new organisms.

that 和 which 均引导定语从句。four important functions 后面的四个 it is 分别介绍土的四种重要作用，modify 改 y 为 i 加 er 表示"调节者、调节器"的意思。that take part in decomposition of organic matter 做 a habitat for organisms 的定语从句。

2. Soil consists of a solid phase (minerals and organic matter) as well as a porous phase that holds gases and water.

as well as 表并列；that holds gases and water 为定语从句，修饰 a porous phase。

3. Accordingly, soils are often treated as a three-state system.

Accordingly 因此，可翻译为"相应地"；be treated as 被当做；three-state system 三相体系，固定短语。

4. be referred to as …

被称为……

3.2 Definition of rock 岩体定义

【Text】

In Geology, rock is a naturally occurring solid aggregate of one or more minerals or mineraloids. For example, the common rock granite is a combination of the quartz, feldspar and biotite minerals. The Earth's outer solid layer, the lithosphere, is made of rock. Fig. 3.2 shows rock outcrop along a mountain creek.

Fig. 3.2 Rock outcrop along a mountain creek
图 3.2 沿山溪的岩石露头

Rocks have been used by mankind throughout history. From the Stone Age rocks have been used for tools. The minerals and metals found in rocks have been essential to human civilization.

Three major groups of rocks are defined: igneous, sedimentary and metamorphic. The scientific study of rocks is called Petrology, which is an essential component of Geology.

Rocks are geologically classified according to characteristics such as mineral and chemical composition, permeability, the texture of the constituent particles, and particle size. These physical properties are the end result of the processes that formed the rocks. Over the course of time, rocks can transform from one type into another, as described by the geological model called the rock cycle. These events produce three general classes of rock: igneous, sedimentary, and metamorphic.

The three classes of rocks are subdivided into many groups. However, there are no hard and fast boundaries between allied rocks. By increase or decrease in the proportions of their constituent minerals they pass by every gradation into one another, the distinctive structures also of one kind of rock may often be traced gradually merging into those of another. Hence the definitions adopted in establishing rock nomenclature merely correspond to more or less arbitrary selected points in a continuously graduated series.

【Key words】

Geology n. 地质学；地质；地理学；地质概况
aggregate n. 聚集体；骨料；集料（可成混凝土或修路等用的）
mineraloid n. 准矿物；似矿物
granite n. 花岗岩，花岗石
quartz n. 石英；水晶；石英石
feldspar n. 长石；长石类
biotite n. 黑云母
Stone Age 石器时代
igneous adj. 火成的；似火的；火成（山）岩的
sedimentary adj. 沉积的，沉淀性的
metamorphic adj. 变质的；变质岩的；变形的
permeability n. 渗透性；渗透率；磁导率；通透性
texture n. 纹理，质地；结构；本质
constituent particle 组成粒子，构成粒子
rock cycle 岩石循环；岩石周期
subdivide v. 细分；再分；分割
allied rock 同源的岩石
gradation n. （从一事物到另一事物的）渐变，层次；分级；渐层；级配
merge into 汇合，（使）并入；归并
nomenclature n. 系统命名法；（某一学科的）术语
arbitrary adj. 随意的，任性的，随心所欲的；主观的

【Translation】

In Geology, rock is a naturally occurring solid aggregate of one or more minerals or mineraloids. For example, the common rock granite is a combination of the quartz, feldspar and biotite minerals. The Earth's outer solid layer, the lithosphere, is made of rock. Fig. 3.2 shows rock outcrop along a mountain creek.

在地质学上,岩石是一种自然状态下存在的一种或多种矿物或似矿物的固体聚集体。例如,常见的花岗岩就是由石英、长石和黑云母矿物集合而成的。地球外部的固体圈层,即岩石圈,就是由岩石构成的。图 3.2 为沿山溪的岩石露头。

Rocks have been used by mankind throughout history. From the Stone Age rocks have been used for tools. The minerals and metals found in rocks have been essential to human civilization.

人类利用岩石的历史十分悠久。从石器时代开始,岩石已经被用作工具。而岩石中发现的矿物和金属对人类文明而言十分重要。

Three major groups of rocks are defined: igneous, sedimentary, and metamorphic. The scientific study of rocks is called petrology, **which is an essential component of** geology.

岩石被分为三大种类:火成岩、沉积岩和变质岩。研究岩石的学科叫做岩理学,是地质学的一个基本组成部分。

Rocks are geologically classified according to characteristics such as mineral and chemical composition, permeability, the texture of the constituent particles, and particle size. These physical properties are the end result of the processes that formed the rocks. Over the course of time, rocks **can transform from one type into another, as described by** the geological model called the rock cycle. These events produce three general classes of rock: igneous, sedimentary, and metamorphic.

在地质学上,岩石根据其特性来分类,比如根据它的矿物和化学成分、渗透率、组成粒子的结构和颗粒大小等。这些物理特性是形成岩石过程的最终结果。随着时间的推移,岩石可以从一种类型转变为另一种类型,就像被称为"岩石循环"的地质模型中描述的那样,从而产生了三种类别的岩石:火成岩、沉积岩和变质岩。

The three classes of rocks are subdivided into many groups. However, there are no hard and fast boundaries between allied rocks. By increase or decrease in the proportions of their constituent minerals they pass by every gradation into one another, the distinctive structures also of one kind of rock may often be traced gradually merging into those of another. Hence the definitions adopted in establishing rock nomenclature merely correspond to more or less arbitrary selected points in a continuously graduated series.

　　三大类岩石可再被细分为许多组别。但是，在类似的岩石之间，并没有说一不二的界限。通过增加或减少岩石中组成矿物的比例，它们可以从一种级别过渡到另一种级别，而且一种岩石的代表性结构常常被发现逐步融合成另一种岩石的结构。因此，在建立岩石名库时所采用的一些定义值仅仅对应于连续渐变数列上的那些随机选择的点。

【Important sentences】
1. The scientific study of rocks is called Petrology, which is an essential component of Geology.
　　which 引导定语从句；an essential component 基本组成部分。
2. Rocks are geologically classified according to characteristics such as mineral and chemical composition, permeability, the texture of the constituent particles, and particle size.
　　be classified according to characteristics 根据特性来分类，geologically 表示从地质学的角度进行的分类。
3. Over the course of time, rocks can transform from one type into another, as described by the geological model called the rock cycle.
　　transform from one type into another 从一种类型转变为另一种类型；as described by…, 就像被……描述的那样。
4. The three classes of rocks are subdivided into many groups.
　　be subdivided into…, 被细分为……

3.3　Mechanical properties of soils　土的力学特性

【Text】
　　The physical properties of soils, in order of decreasing importance, are texture, structure, density, porosity, consistency, temperature, colour and resistivity. Most of these determine the aeration of the soil and the ability of water to infiltrate and to be held in the soil. Soil texture is determined by the relative proportion of the three kinds of soil particles, called soil "separates": sand, silt, and clay. Fig. 3.3 shows soil types by clay, silt and sand composition as used by the USDA. Larger soil structures called "peds" are created from the separates when iron oxides, carbonates, clay, and silica with the organic constituent humus, coat particles and cause them to adhere into larger, relatively stable secondary structures. Soil density, particularly bulk density, is a measure of soil compaction. Soil porosity consists of the part of the soil volume occupied by gases and water.

Soil consistency is the ability of soil to stick together. Soil temperature and colour are self-defining. Resistivity refers to the resistance to conduction of electric currents and affects the rate of corrosion of metal and concrete structures. The properties may vary through the depth of a soil profile.

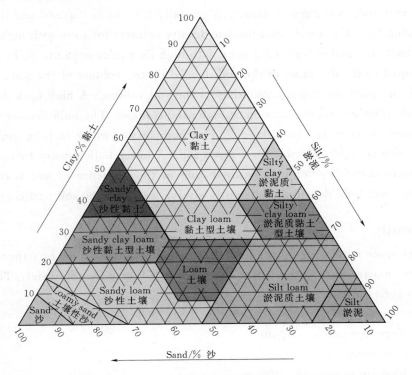

Fig. 3.3 Soil types by clay, silt and sand composition as used by the USDA
图 3.3 美国农业部所用的根据黏土、淤泥和沙组成进行的土壤分类

Texture

The mineral components of soil are sand, silt and clay, and their relative proportions determine a soil's texture.

Structure

The clumping of the soil textural components of sand, silt and clay forms aggregates and the further association of those aggregates into larger units forms soil structures called peds. Soil structure often gives clues to its texture, organic matter content, biological activity, past soil evolution, human use, and the chemical and mineralogical conditions under which the soil formed. While texture is defined by the mineral component of a soil and is an innate property of the soil that does not change with agricultural activities, soil structure can be improved or destroyed by the choice and timing of farming practices.

Density

Density is the weight per unit volume of an object. Particle density is equal to the mass of solid particles divided by the volume of solid particles—it is the density of only the mineral particles that make up a soil; i.e., it excludes pore space and organic material. Soil particle density is typically 2.60 to 2.75g/cm^3 and is usually unchanging for a given soil. Soil particle density is lower for soils with high organic matter content, and is higher for soils with high Fe-oxides content. Soil bulk density is equal to the dry mass of the soil divided by the volume of the soil; i.e., it includes air space and organic materials of the soil volume. A high bulk density is indicative of either soil compaction or high sand content. The bulk density of cultivated loam is about 1.1 to 1.4g/cm^3 (for comparison water is 1.0g/cm^3). Soil bulk density is highly variable for a given soil. A lower bulk density by itself does not indicate suitability for plant growth due to the influence of soil texture and structure. Soil bulk density is inherently always less than the soil particle density.

Porosity

Pore space is that part of the bulk volume that is not occupied by either mineral or organic matter but is open space occupied by either gases or water. There are four categories of pores:

- Very fine pores: <2μm
- Fine pores: 2 – 20μm
- Medium pores: 20 – 200μm
- Coarse pores: 200μm – 0.2mm

Consistency

Consistency is the ability of soil to stick to itself or to other objects (cohesion and adhesion respectively) and its ability to resist deformation and rupture. It is of rough use in predicting cultivation problems and the engineering of foundations.

The terms used to describe the soil consistency in three moisture states and a last consistency not affected by the amount of moisture are as follows:

- Consistency of Dry soil: loose, soft, slightly hard, hard, very hard, extremely hard
- Consistency of moist soil: loose, very friable, friable, firm, very firm, extremely firm
- Consistency of wet soil: nonsticky, slightly sticky, sticky, very sticky; nonplastic, slightly plastic, plastic, very plastic
- Consistency of cemented soil: weakly cemented, strongly cemented, indurated

(requires hammer blows to break up)

Soil consistency is useful in estimating the ability of soil to support buildings and roads. More precise measures of soil strength are often made prior to construction.

Temperature

Soil temperature depends on the ratio of the energy absorbed to that lost. Soil has a temperature range from −20 to 60℃. Soil temperature regulates seed germination, plant and root growth and the availability of nutrients.

There are various factors that affect soil temperature, such as water content, soil color, relief (slope, orientation, and elevation), and soil cover (shading and insulation).

Color

Soil color is often the first impression one has when viewing soil. Striking colors contrasting patterns are especially noticeable. The Red River (Mississippi watershed) carries sediment eroded from extensive reddish soils like port silt loam in Oklahoma. The Yellow River in China carries yellow sediment from eroding loess soils. Mollisols in the Great Plains of North America are darkened and enriched by organic matter. Podsols in boreal forests have highly contrasting layers due to acidity and leaching.

In general, color is determined by the organic matter content, drainage conditions, and degree of oxidation. Soil color, while easily discerned, has little use in predicting soil characteristics.

Resistivity

Soil resistivity is a measure of soil's ability to retard the conduction of an electric current. The electrical resistivity of soil can affect the rate of galvanic corrosion of metallic structures in contact with the soil. Higher moisture content or increased electrolyte concentration can lower resistivity and increase conductivity, thereby increasing the rate of corrosion.

【Key words】

density *n.* 密度；浓度；比重
porosity *n.* 孔隙率；多孔性；孔隙度
consistency *n.* 一致性；连贯性，连续性
resistivity *n.* 电阻率；电阻系数
aeration *n.* 通风；换气；松砂

infiltrate v. （使）渗透，（使）渗入
particle n. 微粒，颗粒
silt n. 淤泥，泥沙，粉砂
clay n. 黏土，泥土
peds n. 土壤结构自然体
oxide n. 氧化物；氧化层
carbonate n. 碳酸盐；碳酸酯；碳酸盐岩
silica n. 石英；硅石；二氧化硅
humus n. 腐殖质；腐殖土
bulk n. 体积；大块，大量；大多数，大部分；主体
compaction n. 压实；压紧
volume n. 体积；音量；容量
conduction n. 传导，导热，导电；传导率
corrosion n. 腐蚀；侵蚀；锈蚀；溶蚀
concrete n. 混凝土；具体物
clumping n. （土、泥等）团；块；聚丛；成块
mineralogical adj. 矿物学的
innate adj. 天生的；先天的；固有的
pore n. 毛孔；气孔；孔隙；细孔
Fe-oxide 铁氧化物
indicative adj. 指示的；陈述的；暗示的
suitability n. 适用性；适宜性；适合性
inherently adv. 固有地；内在地；本质上
category n. 类别；分类；种类
fine adj. 上等的；纤细的，精致的
coarse adj. 粗糙的；粗的；粗劣的
adhesion n. 粘连；黏合；[物] 黏附力
deformation n. 金属等（在压力作用下）变形，形变
rupture n. 破裂；断裂；决裂；裂开
friable adj. 脆的，易碎的
firm adj. 坚固的，坚牢的；坚定的
nonsticky adj. （=non-sticky）无黏性的；不黏的
nonplastic adj. 非可塑性的；非塑性的
cemented adj. 注水泥的；胶结的
indurated adj. 硬化的；硬结的；固化的；固结的
ratio n. 比，比率；比例；系数
regulate v. 控制；管理；调节；调整
germination n. 萌芽，发生；萌发
nutrient n. 营养物，营养品，养分
shading n. 着色；描影；明暗；色差

insulation *n*. 绝缘；隔离；隔热；孤立
contrasting pattern 对比格局
watershed *n*. 分水岭；流域；集水区；分水线
erode *v*. 腐蚀；侵蚀；腐蚀侵蚀；冲刷
reddish *adj*. 微红的；淡红；带红色的
acidity *n*. 酸度；酸味；酸值；酸碱度
leach *v*. 使（液体）过滤；滤去某物中的（可溶物质）
drainage *n*. 排水，放水；排水系统；排走物，废水
oxidation *n*. 氧化；氧化作用；氧化反应
discern *v*. 辨别；辨明；分辨；洞悉
galvanic *adj*. 流电的，抽搐的
metallic *adj*. 金属的；金属性的；金属制的

【Translation】

The physical properties of soils, in order of decreasing importance, are texture, structure, density, porosity, consistency, temperature, colour and resistivity. Most of these determine the aeration of the soil and the ability of water to infiltrate and to be held in the soil. Soil texture is determined by the relative proportion of the three kinds of soil particles, called soil "separates": sand, silt, and clay. Fig. 3.3 shows soil types by clay, silt and sand composition as used by the USDA. when iron oxides, carbonates, clay, and silica with the organic constituent humus, coat particles and cause them to adhere into larger, relatively stable secondary structures, larger soil structures called "peds" are created from the separates. Soil density, particularly bulk density, is a measure of soil compaction. Soil porosity **consists of** the part of the soil volume occupied by gases and water. Soil consistency is the ability of soil to stick together. Soil temperature and colour are self-defining. Resistivity refers to the resistance to conduction of electric currents and affects the rate of corrosion of metal and concrete structures. The properties may vary through the depth of a soil profile.

按照重要性递减排序，土体的物理性质依次为结构、构造、密度、孔隙度、连续性、温度、颜色和电阻率。这些参数的大部分决定了土壤的通气性和渗水及储水能力。土的结构由三种土粒子的相对比例决定，即称为土壤"分散体"的砂、粉砂和黏土。图3.3为美国农业部所用的根据黏土、粉砂和砂组成进行土壤分类的模型。当含有有机组成腐殖质的铁的氧化物、碳酸盐岩、黏土和二氧化硅等将颗粒包裹起来，并使它们黏结在一起变成更大、相对稳定的次生构造时，便由"分散体"形成了被称为"土块"的较大的土壤构造。而土壤密度，特别是体积密度，是土壤压实度的测量指标。土壤孔隙度由土壤中被气体和水所占的那部分体积组成。土壤的黏连性则是反映土体黏结在一起的能力。土的温度和颜色是自定义参数。电阻率指电流传导的阻力，它影响金属和混凝土构件的腐蚀速率。这些特性可能随土体剖

面深度的变化而变化。

Texture　结构

The mineral components of soil are sand, silt and clay, and their relative proportions determine a soil's texture.

土壤的矿物成分是砂、粉砂和黏土，它们的相对比例决定了土壤的结构。

Structure　构造

The clumping of the soil textural components of sand, silt and clay forms aggregates and the further association of those aggregates into larger units forms soil structures called peds. Soil structure often gives clues to its texture, organic matter content, biological activity, past soil evolution, human use, and the chemical and mineralogical conditions under which the soil formed. **While texture is defined by** the mineral component of a soil and is an innate property of the soil that **does not change with** agricultural activities, soil structure **can be improved or destroyed by** the choice and **timing** of farming practices.

土的组成成分砂、粉砂和黏土结成块就形成了聚集体，这些聚集体进一步聚合成更大的单元体并形成称为土壤自然结构体的土壤构造。土壤构造往往能反映关于土壤的结构、有机质成分、生物活性、过往土的变迁、人类使用以及土壤在形成时的化学和成矿条件等情况。土壤结构是由土壤的矿物成分定义的，因此是一种不随农耕活动而改变的固有特性，但土的构造则可以通过农事活动的选择和时机把握来获得改良或导致破坏。

Density　密度

Density is the weight per unit volume of an object. Particle density **is equal to** the mass of solid particles **divided by** the volume of solid particles—**it is** the density of only the mineral particles **that** make up a soil; i.e., it excludes pore space and organic material. Soil particle density is typically 2.60 to 2.75g/cm^3 and is usually unchanging for a given soil. Soil particle density is lower for soils with high organic matter content, and is higher for soils with high Fe-oxides content. Soil bulk density is equal to the dry mass of the soil divided by the volume of the soil; i.e., it includes air space and organic materials of the soil volume. A high bulk density is indicative of either soil compaction or high sand content. The bulk density of cultivated loam is about 1.1 to 1.4g/cm^3 (for comparison water is 1.0g/cm^3). Soil bulk density is highly variable for a given soil. A lower bulk density by itself does not indicate suitability for plant growth due to the influence of soil texture and structure. Soil bulk density is inherently always less than the soil particle density.

密度是物体单位体积的重量。颗粒密度等于固体颗粒的质量除以固体颗粒的体积——它只是构成土壤的矿物颗粒的密度;也就是说,它不包括孔隙和有机质。土体颗粒密度通常是 $2.60\sim2.75g/cm^3$,而且对于给定的土壤其值通常是不变的。有机质含量高的土壤,其颗粒密度较低,而铁氧化物含量高的土壤,其颗粒密度较高。土壤的体积密度等于土壤干燥质量除以土壤的体积;也就是说,它包括了空气所占空间和土中有机质材料的体积。高体积密度表明土壤压实程度高或含砂量高。栽培用土壤的体积密度约为 $1.1\sim1.4g/cm^3$(参照物水为 $1g/cm^3$)。对于给定的土壤种类,土壤体积密度变化极大。本身体积密度低并不能说明其对植物生长的适宜性,因为要考虑土壤结构和构造的影响。土壤的体积密度本质上总是小于土壤的颗粒密度。

Porosity 孔隙度

Pore space is that part of the bulk volume that is not occupied by either mineral or organic matter but is open space occupied by either gases or water. There are four categories of pores:

孔隙空间是指不被矿物或有机质所占据的那一部分体积,它是由气体或水所占据的开放空间。孔隙有 4 种级别:

- Very fine pores: $<2\mu m$
  极细孔隙: $<2\mu m$
- Fine pores: $2-20\mu m$
  细孔隙: $2\sim20\mu m$
- Medium pores: $20-200\mu m$
  中等孔隙: $20\sim200\mu m$
- Coarse pores: $200\mu m-0.2mm$
  粗孔隙: $200\mu m\sim0.2mm$

Consistency 粘连性

Consistency is the ability of soil to stick to itself or to other objects (cohesion and adhesion respectively) and its ability to resist deformation and rupture. It is of rough use in predicting cultivation problems and the engineering of foundations.

粘连性是土壤对自身或其他物体黏结的能力(分别是黏聚力和黏附力),以及抵抗变形和断裂的能力。它可以用于粗略预测耕种问题以及基础工程。

The terms used to describe the soil consistency in three moisture states and a last consistency not affected by the amount of moisture are as follows:

下面是用于描述三种不同湿度状态下土壤粘连性的术语,最后一项粘连性不受湿度的影响：

- Consistency of dry soil: loose, soft, slightly hard, hard, very hard, extremely hard
  干土粘连性：松散，松软，微硬，硬，坚硬，很坚硬
- Consistency of moist soil: loose, very friable, friable, firm, very firm, extremely firm
  潮土粘连性：松散，很易碎，易碎，坚固，很坚固，非常坚固
- Consistency of wet soil: nonsticky, slightly sticky, sticky, very sticky; nonplastic, slightly plastic, plastic, very plastic
  湿土粘连性：不粘，稍粘，粘的，很粘；不可塑，稍可塑，可塑，易塑
- Consistency of cemented soil: weakly cemented, strongly cemented, indurated (requires hammer blows to break up)
  胶结土粘连性：弱胶结，强胶结，硬结的（需要捶打才能破坏）

Soil consistency is useful in estimating the ability of soil to support buildings and roads. More precise measures of soil strength are often made prior to construction.

土壤的粘连性在估算土壤支撑建筑物和道路的能力方面是很有用的。而施工之前，往往需要对土体强度进行更精确的测量。

Temperature 温度

Soil temperature depends on the ratio of the energy absorbed to that lost. Soil has a temperature range from -20 to $60℃$. Soil temperature regulates seed germination, plant and root growth and the availability of nutrients.

土壤温度取决于所吸收的能量与损失的能量之比。土壤的温度范围介于-20~$60℃$之间。土壤温度调控种子发芽、植株和根系生长以及营养物的获取。

There are various factors that affect soil temperature, such as water content, soil color, relief (slope, orientation and elevation) and soil cover (shading and insulation).

存在各种影响土壤温度的因素，比如含水率、土壤颜色和地形（坡度、方向、标高），还有土壤覆盖情况（遮光和保温性）。

Color 颜色

Soil color is often the first impression one has when viewing soil. Striking colors and contrasting patterns are especially noticeable. The Red River (Mississippi watershed) carries sediment eroded from extensive reddish soils like port silt loam in Oklahoma. The Yellow River in China carries yellow sediment from eroding loess soils. Mollisols in the Great Plains of North America are darkened and enriched by organic matter. Podsols in boreal forests have highly contrasting layers due to acidity and leaching.

土壤的颜色经常是人们看到土壤时会有的第一印象。醒目的颜色和有反差的图案特别引人注意。红河（密西西比河流域）搬运从像俄克拉荷马港淤泥土壤一样的红土上侵蚀而来的沉积物。而中国的黄河会携带因侵蚀黄土而来的黄色沉积物。在北美洲大平原上的黑土因为有机质而颜色加深并变得肥沃。北方森林的灰化土因为酸性和浸出作用会形成高度反差层。

In general, color is determined by the organic matter content, drainage conditions, and degree of oxidation. Soil color, while easily discerned, has little use in predicting soil characteristics.

总的来说，土壤的颜色是由有机质含量、排水条件和氧化程度来决定的。尽管土壤的颜色容易辨认，但是在预测土壤特性方面用得很少。

Resistivity　电阻率

Soil resistivity is a measure of soil's ability to retard the conduction of an electric current. The electrical resistivity of soil can affect the rate of galvanic corrosion of metallic structures in contact with the soil. Higher moisture content or increased electrolyte concentration can lower resistivity and increase conductivity, thereby increasing the rate of corrosion.

土壤电阻率是衡量土壤阻碍电流传导能力的一种量度。土壤的电阻率可以影响与土壤接触的金属结构的电化学腐蚀速率。较高的含水量或增加电解质浓度可以降低电阻率并提高导电率，从而提高腐蚀速率。

【Important sentences】
1. Soil porosity consists of the part of the soil volume occupied by gases and water. consist of… 由……组成。
2. While texture is defined by the mineral component of a soil and is an innate property of the soil that does not change with agricultural activities, soil structure can be improved or destroyed by the choice and timing of farming practices. While 表转折；be defined by… 通过……来进行定义；do not change with… 不会随着……而发生改变，此处的 that 从句修饰 an innate property；be improved or destroyed by… 通过……来进行改良或导致破坏；timing 为动名词，表时机。
3. Particle density is equal to the mass of solid particles divided by the volume of solid particles—it is the density of only the mineral particles that make up a soil; i. e., it excludes pore space and organic material.

be equal to 等于; divided by 除以; "—" 进行解释说明; it is the density of only the mineral particles that make up a soil 是强调句型。

3.4 Mechanical properties of rocks 岩体的力学特性

【Text】

Information collected by geologists and engineering geologists is in general not sufficient to predict the engineering behavior of rocks and rock masses. Tests need to be conducted to assess the response of rocks under a wide variety of disturbances such as static and dynamic loading, seepage and gravity and the effect of atmospheric conditions and applied temperatures. In general, rock and rock mass properties can be divided into five groups:

- physical properties (durability, hardness, porosity, etc.)
- mechanical properties (deformability, strength)
- hydraulic properties (permeability, storativity)
- thermal properties (thermal expansion, conductivity)
- in-situ stresses

When exposed to atmospheric conditions, rocks slowly break down. This process is called weathering and can be separated into mechanical (also called physical) weathering and chemical weathering.

Mechanical weathering causes disintegration of rocks into smaller pieces by exfoliation or decrepitation (slaking). The chemical composition of the parent rock is not or is only slightly altered. Mechanical weathering can result from the action of agents such as frost action, salt crystallization, temperature changes (freezing and thawing), moisture changes (cycles of wetting and drying), wind, glaciers, streams, unloading of rock masses (sheet jointing), and biogenic processes (plants, animals, etc.). For instance, mechanical weathering is very active in high mountains with cold climates.

Chemical weathering creates new minerals in place of the ones it destroys in the parent rock. As rocks are exposed to atmospheric conditions at or near the ground surface, they react with components of the atmosphere to form new minerals. The most important atmospheric reactants are oxygen, carbon dioxide, and water. In polluted air, other reactants are available (acid rain problems associated with the release of sulfuric acid from coal-fired power plants, sulfur dioxide and smoke emissions, nitrogen oxides from vehicle exhaust). In general, chemical weathering reactions are exothermic and cause volume increases.

Solution is a reaction whereby a mineral completely dissolves during weathering. This type of reaction depends on the solubility of the rock minerals. For instance, evaporite minerals (salt, gypsum) dissolve quickly in water, whereas carbonate minerals are somewhat less soluble.

Limestone dissolves by meteoric water which contains dissolved carbon dioxide. This results in the formation of cavities called dolines or karsts and geologic hazards called sink holes.

【Key words】

disturbance *n.* 干扰；扰动；动乱
static *n.* 静态；静态的；静止的
dynamic *n.* 动态；动力　*adj.* 动态的
seepage *n.* 渗流；渗透；渗漏；浸润
durability *n.* 耐用性；耐久性；持久性
mechanical *adj.* 机械的；力学的
deformability *n.* 变形性；可变形性；变形能力
hydraulic *adj.* 液压的；水力的
storativity *n.* 储存系数；储水系数
in-situ *n.* 原位；现场　*adj.* 原位的
disintegration *n.* 蜕变；瓦解；衰变；分解
exfoliation *n.* 剥落；脱落；表皮脱落
decrepitation *n.* 爆裂；爆裂作用
slake *v.* 水化；潮解；销蚀
frost action 冻裂作用；冰冻作用
crystallization *n.* 结晶作用；结晶化；结晶过程
sheet jointing 页状剥落
biogenic *adj.* 源于生物的，生物所造成的
reactant *n.* 反应物；成分；反应剂
sulfuric *adj.* 硫的，含（六价）硫的
exothermic *adj.* 放热的；发热的；产热的
solution *n.* 溶液，溶解
solubility *n.* 溶解度；可溶性；溶解性；溶度
evaporate *v.* 消失；挥发；使蒸发，使脱水
gypsum *n.* 石膏；生石膏；硫酸钙
soluble *adj.* 可溶的；可以解决的
limestone *n.* 石灰石；石灰岩
meteoric *adj.* 疾速的；令人眼花缭乱的

cavity *n.* 空洞；腔；洞
doline *n.* 灰岩坑；落水洞；溶斗
karst *n.* 喀斯特；岩溶；喀斯特地形（地貌）
sink hole 灰岩坑；地洞；缩孔

【Translation】

Information collected by geologists and engineering geologists is in general not sufficient to predict the engineering behavior of rocks and rock masses. **Tests need to be conducted to assess** the response of rocks under a wide variety of disturbances such as static and dynamic loading, seepage and gravity and the effect of atmospheric conditions and applied temperatures. In general, rock and rock mass properties can be divided into five groups:

地质学家和工程地质学家收集的信息通常不足以预测岩石和岩体的工程特性。需要开展一些实验对岩石在各种扰动下的力学响应进行评估，如静态和动态加载、渗流、重力和大气条件的影响，以及所施加的温度等。一般来说，岩石和岩体性质可以分为五类：

- physical properties (durability, hardness, porosity, etc.)
  物理性质（耐久性、硬度、孔隙率等）
- mechanical properties (deformability, strength)
  力学性质（变形性、强度）
- hydraulic properties (permeability, storativity)
  水力学特性（渗透性、储水系数）
- thermal properties (thermal expansion, conductivity)
  温度特性（热膨胀、导电性）
- in-situ stresses
  现场应力

When exposed to atmospheric conditions, rocks slowly break down. This process is called weathering and can be separated into mechanical (also called physical) weathering and chemical weathering.

当暴露在大气条件下时，岩石会慢慢分解。这一过程被称为风化作用，并且可分为机械（也称物理）风化和化学风化。

Mechanical weathering causes disintegration of rocks into smaller pieces by exfoliation or decrepitation (slaking). The chemical composition of the parent rock is not or is only slightly altered. Mechanical weathering can **result from** the action of agents such as frost action, salt crystallization, temperature changes (freezing and thawing), moisture

3.4 Mechanical properties of rocks 岩体的力学特性

changes (cycles of wetting and drying), wind, glaciers, streams, unloading of rock masses (sheet jointing), and biogenic processes (plants, animals, etc.). For instance, mechanical weathering is very active in high mountains with cold climate.

机械风化作用使岩石因为剥落或爆裂（崩解）瓦解成较小的碎块。母岩的化学成分不变或只是稍微有一点改变。机械风化可能是因为像冻结作用、盐结晶作用、温度变化（冻结和融化）、湿度变化（干湿循环）、风、冰川、溪流、岩体卸荷（页状剥落）和生物过程（植物、动物等）等因素的作用引起的结果。例如，机械风化作用在寒冷气候的高山地区是非常活跃的。

Chemical Weathering creates new minerals in place of the ones it destroys in the parent rock. As rocks are exposed to atmospheric conditions at or near the ground surface, they react with components of the atmosphere to form new minerals. The most important atmospheric reactants are oxygen, carbon dioxide, and water. In polluted air, other reactants are available (acid rain problems associated with the release of sulfuric acid from coal-fired power plants, sulfur dioxide and smoke emissions, nitrogen oxides from vehicle exhaust). In general, chemical weathering reactions are exothermic and cause volume increases.

化学风化会生成新的矿物取代母岩中被破坏的矿物。当地表或临近地表的岩石暴露在大气条件下时，它们与大气中的成分发生反应而形成新的矿物。大气中最重要的反应物是氧气、二氧化碳和水。在受污染的空气中，存在其他反应物（与从燃煤火力发电厂排放的硫酸气体相关的酸雨问题，二氧化硫和烟雾的排放，来自汽车尾气的氮氧化物）。一般情况下，化学风化反应是放热的，并导致体积增大。

Solution is a reaction whereby a mineral completely dissolves during weathering. This type of reaction depends on the solubility of the rock minerals. For instance, evaporite minerals (salt, gypsum) dissolve quickly in water, whereas carbonate minerals are somewhat less soluble.

溶解是一种矿物在风化过程中完全分解的反应。这种反应取决于岩石矿物的溶解度。例如，蒸发岩矿物（盐类、石膏）可迅速地溶解在水中，而碳酸盐矿物不是那么好溶解的。

Limestone dissolves by meteoric water **which** contains **dissolved** carbon dioxide. This results in the formation of cavities called dolines or karsts and geologic hazards called sink holes.

石灰岩可被含有溶解的二氧化碳的雨水所溶解，由此形成被称为漏斗或卡斯特溶洞的洞穴以及称为塌陷坑的地质灾害。

【Important sentences】

1. Information collected by geologists and engineering geologists is in general not sufficient to predict the engineering behavior of rocks and rock masses.
   Information collected by geologists and engineering geologists 由地质学家和工程地质学家收集的信息。collected，过去分词作定语；be in general not sufficient to 表示通常不足以做某事，in general 表通常。
2. Tests need to be conducted to assess…
   被动句，需要开展一些实验去评估……
3. Mechanical weathering can result from the action of agents such as…
   result from 起因于。
4. Limestone dissolves by meteoric water which contains dissolved carbon dioxide.
   dissolves by 发生溶解；which 修饰 meteoric water，引导定语从句；dissolved 表示溶解的，过去分词作定语。

3.5 Testing methods and procedures 测试方法与流程

【Text】

The ability to apply different stress paths and boundary conditions to a given sample or suite of samples allows different failure modes to be investigated in the laboratory. To determine the appropriate parameters that describe the rock behavior the following tests are commonly performed.

Unconfined and confined compression tests

For example, the uniaxial test is the most frequently used rock mechanics test, but provides only elastic properties and a single failure value derived from a very simple stress path.

With computer controlled feedback it is possible to follow different stress paths by varying the axisymmetric confining pressure and the axial compression of a rock cylinder.

Triaxial tests are done to measure the rock strength as a function of confining pressure. The testing procedure is to maintain both confining pressure and internal axial load for two hours. After maintaining a base line, the sample is loaded axially by increasing the axial stress with constant strain rates at 0.006m/s. At the onset of failure, the confining pressure is maintained, and the axial loading continues until failure occurs. Expected duration of the test is approximately twelve hours. Axial, radial, external load, internal load, and confining pressure data will be provided along with a report. Triaxial multiple state test results and analyses are shown in Fig. 3.4.

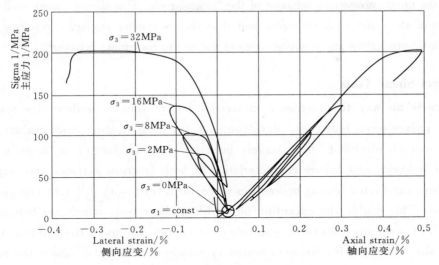

(a) Relationships between stresses and strains
应力-应变关系

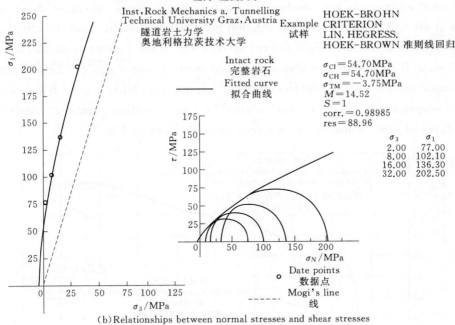

(b) Relationships between normal stresses and shear stresses
法向应力与剪应力关系

Fig. 3.4 Triaxial multiple state test results and analyses
图 3.4 三轴多状态试验结果及分析

The use of computed automated controls allows us to perform multiple loading cycles on the same specimen. After each peak load for a given confining pressure the deviatoric stress is reduced to zero and the sample is loaded hydrostatically to the next confining level. Thus, the progressive stress history of a single sample can be monitored instead of using different samples (In fact, different samples may have different microstructure) at each stress state and combining the results to estimate the progressive stress behavior of the "intact" rock. This allows a more realistic evaluation of the intact rock strength, and thus the rock mass strength, resulting in more realistic predictions and interpretations of the in-situ rock mass behavior.

Direct Shear Tests

There is no way to describe a material's failure criteria without its shear strength parameters. However, developments in shear testing and evaluation methods in rock mechanics have largely been ignored. Therefore, to investigate the shear behavior and failure characteristics of both fracture surfaces and intact rock we use automated testing procedures to perform tests with different boundary conditions. This enable the execution of modified shear tests which are behavior specific. For example, stiffness controlled tests can be used to evaluate the ultimate shear strength for different boundary conditions, and also allows the recognition of the different failure modes that occur during shearing. This test method is the most appropriate test method for evaluating a material's shear behavior such as the shear and normal stiffness, dilation potential, cohesion, and the initial and ultimate friction angles. Multi-failure state shear tests (under constant normal loads) as well as various combinations of test control procedures can be performed on a single sample eliminating the effects of sample variability on the failure envelope. Comparison between constant normal load and constant normal stiffness test procedures of direct shear tests on a rock joint is shown in Fig. 3.5.

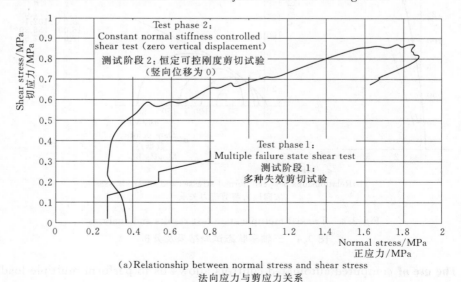

(a) Relationship between normal stress and shear stress
法向应力与剪应力关系

Fig. 3.5 (1) Comparison between constant normal load and constant normal stiffness test procedures of direct shear tests on a rock joint
图 3.5 (一) 岩石节理直剪试验中法向荷载不变与法向刚度不变试验流程对比

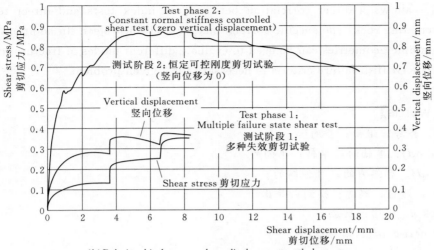

(b) Relationship between shear displacement and shear stress
剪应力与剪切位移关系

Fig. 3.5 (2) Comparison between constant normal load and constant normal stiffness test procedures of direct shear tests on a rock joint
图 3.5（二） 岩石节理直剪试验中法向荷载不变与法向刚度不变试验流程对比

Tension Tests

The preparation of end performing tests for a direct evaluation of the tensile strength is difficult and not widely used. Instead many index tests such as the Brazilian test, three or four point bending tests, etc. are used. The correlation between the test results and the direct tensile strength is not often clear since different stress paths are responsible for the rock failure. Using overcored samples we test the direct tensile strength of a sample using compressional loading (Fig. 3.6).

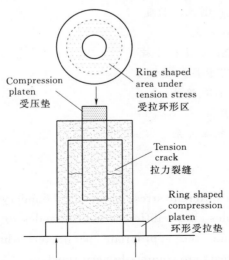

Fig. 3.6 Test the direct tensile strength of a sample using compressional loading
图 3.6 使用加压的方法测定试样的直接拉伸强度

This allows a direct correlation between the index tests and direct tensile values creating rock type specific correlations for a given project. In addition, different sample geometries can be used to evaluate different modes of fracture toughness if the Griffith failure criteria is necessary for a given problem.

【Key words】

suite *n*. 套房；组曲；套间
uniaxial *adj*. 单轴的；单轴；单向的
elastic *adj*. 有弹力的；可伸缩的
derive *v*. 源自；起源；得到；导出
feedback *n*. 反馈；反应
axisymmetric *adj*. 轴对称的
cylinder *n*. 圆柱体；汽缸；圆筒
triaxial *adj*. 三轴的，三维的，空间的
maintain *v*. 维持；保持
radial *adj*. 辐射状的；放射式的；径向的
snapshot *n*. 快照；简介；急射
automated *adj*. 自动化的；自动的
specimen *n*. 标本；样品；样本；试样
deviatoric *n*. 偏应力
hydrostatical *adj*. 静水力学的，流体静力学的
progressive stress 递增应力
intact *adj*. 完整的；完好的；完整无缺的
parameter *n*. 参数；参量；指标参数
fracture *n*. 破裂，断裂；骨折
stiffness *n*. 硬度；刚度；刚性；劲度
dilation *n*. 膨胀，扩张，扩大
eliminate *v*. 排除，消除；淘汰；除掉
variability *n*. 变异性；可变性；变化性
envelope *n*. 封套；封皮；包络线
index *n*. 索引；指数；指标
correlation *n*. 相关性；相互关系

【Translation】

The ability to apply different stress paths and boundary conditions **to** a given sample or suite of samples allows different failure modes to be investigated in the laboratory. To determine the appropriate parameters that describe the rock behavior the following tests are commonly performed.

能够对给定的试样或一套试样施加不同的应力路径和边界条件的能力使得在实

3.5 Testing methods and procedures 测试方法与流程

验室对不同的破坏模式进行调查成为可能。为了确定用于描述岩石特性的适当参数，通常可进行下面一些试验。

Unconfined and confined compression tests 无围压和有围压抗压试验

For example, the uniaxial test is the most frequently used rock mechanics test, but provides only elastic properties and a single failure value derived from a very simple stress path.

比如，单轴试验是最常用的岩石力学试验，但只能提供弹性性能参数和由很简单的应力路径所造成的单一破坏值。

With computer controlled feedback it is possible to follow different stress paths by varying the axisymmetric confining pressure and the axial compression of a rock cylinder.

随着计算机反馈控制技术的发展，使得能够通过改变轴对称围压和岩柱轴向压力的方法跟踪不同的应力路径。

Triaxial tests are done to measure the rock strength as a function of confining pressure. The testing procedure is to maintain both confining pressure and internal axial load for two hours. After maintaining a base line, the sample is loaded axially by increasing the axial stress with constant strain rates at 0.006m/s. At the onset of failure, the confining pressure is maintained, and the axial loading continues until failure occurs. Expected duration of the test is approximately twelve hours. Axial, radial, external load, internal load, and confining pressure data will be provided along with a report. Triaxial multiple state test results and analyses are shown in Fig. 3.4.

三轴试验可用来测量岩石强度与围压的函数关系。测试流程是，先将围压和内部轴压保持两个小时。在获得一个基准线以后，通过保持0.006mm/s恒应变率的方式增加轴向应力对岩样进行轴向加载。在破坏开始时，围压保持不变，继续增加轴向载荷直到破坏发生。预计试验时间约为12h。将以试验报告形式提供轴向、径向、外部荷载，内部荷载和围压等数据。三轴多状态试验结果及分析如图 3.4所示。

The use of computed automated controls allows us to perform multiple loading cycles on the same specimen. After each peak load for a given confining pressure the deviatoric stress is reduced to zero and the sample is loaded hydrostatically to the next confining level. Thus, the progressive stress history of a single sample **can be monitored instead of** using different samples (with different microstructure?)

at each stress state and combining the results to estimate the progressive stress behavior of the "intact" rock. This allows a more realistic evaluation of the intact rock strength, and thus the rock mass strength, resulting in more realistic predictions and interpretations of the in-situ rock mass behavior.

计算机自动控制的运用使我们能够在同一个试样上进行多载荷循环。对于给定的围压,当峰值荷载达到后,就将偏应力减少至零,并使试样以静水压力方式加载至下一个围压水平。因此,能够监测单一试样上的应力渐进过程,而不是在每一个应力阶段都使用不同的试样(可能具有不同的微观结构?),而后将结果组合起来预测"致密"岩石对渐进应力的响应特性。这样可以得到对致密岩石强度更客观的评价,并获得更客观的岩体强度,使得对现场岩体特性的预测和解释更符合实际。

Direct Shear Tests 直接剪切试验

There is no way to describe a material's failure criteria **without** its shear strength parameters. However, developments in shear testing and evaluation methods in rock mechanics have largely been ignored. Therefore, to investigate the shear behavior and failure characteristics of both fracture surfaces and intact rock we use automated testing procedures to perform tests with different boundary conditions. This enable the execution of modified shear tests which are behavior specific. For example, stiffness controlled tests can be used to evaluate the ultimate shear strength for different boundary conditions, and also allows the recognition of the different failure modes that occur during shearing. This test method is the most appropriate test method for evaluating a material's shear behavior such as the shear and normal stiffness, dilation potential, cohesion, and the initial and ultimate friction angles. Multi-failure state shear tests (under constant normal loads) as well as various combinations of test control procedures can be performed on a single sample eliminating the effects of sample variability on the failure envelope. Comparison between constant normal load and constant normal stiffness test procedures of direct shear tests on a rock joint is shown in Fig. 3.5.

如果没有材料的剪切强度参数,就无法描述材料的破坏准则。但是,岩石力学领域关于剪切试验和评价方法的发展已经被严重忽视了。因此,为了研究断裂面和完整岩石的剪切性能和破坏特征,我们使用自动化测试程序来进行具有不同边界条件的试验。例如,刚度控制试验可以用来评估不同边界条件下的极限剪切强度,并且还能用于对发生在剪切过程中的不同破坏模式进行识别。这种试验方法是用于评估材料剪切特性最适合的试验方法,如切向和法向刚度、扩张势能、内聚力及初始和极限摩擦角等。多破坏阶段的剪切试验(在法向荷载恒定条件下)以及各种组合的流程控制试验能够针对单一试样来完成,从而避免在绘制破坏包络线时出现因试样改变所带来的影响。岩石节理直剪试验中法向荷载不变与法向刚度不变试验流程的对比如图 3.5 所示。

Tension Tests　拉伸试验

The preparation of performing tests for a direct evaluation of the tensile strength is difficult and not widely used. Instead many index tests such as the Brazilian test, three or four point bending tests, etc. are used. The correlation between the test results and the direct tensile strength is not often clear since different stress paths are responsible for the rock failure. Using overcored samples we test the direct tensile strength of a sample using compressional loading. This allows a direct correlation between the index tests and direct tensile values creating rock type specific correlations for a given project. In addition, different sample geometries can be used to evaluate different modes of fracture toughness if the Griffith failure criteria is necessary for a given problem.

能用于对拉伸强度进行直接评价的试验准备起来是困难的，也不被广泛采用。相反，会使用许多指数性试验，如巴西劈裂试验、三点或四点弯曲试验等。测试结果和直接拉伸强度之间的相关性通常不是很明显，因为是不同的应力路径造成岩石发生破坏。借助一种去芯的试样，可以通过加压的方式测试试样的直接拉伸强度。这样能够获得指数性测试与直接拉伸值之间的直接关系，形成针对给定项目的岩石类型特定相关关系。此外，如果针对给定的问题需要用到格里菲思破坏准则时，可以使用不同的试样几何形状来评估不同类型的断裂强度。

【Important sentences】

1. The ability to apply different stress paths and boundary conditions to a given sample or suite of samples allows different failure modes to be investigated in the laboratory.
   句子的主干部分为：The ability allows different failure modes to be investigated in the laboratory。第一个 to 后面的 "apply different stress paths and boundary conditions to a given sample or suite of samples（在给定的试样或一套试样施加不同的应力路径和边界条件）" 作 The ability 的定语。
2. Thus, the progressive stress history of a single sample can be monitored instead of using different samples…
   can be monitored 表被动；instead of 表转折，翻译为"而不是"。
3. There is no way to describe a material's failure criteria without its shear strength parameters.
   There is no way to do sth. without…，固定表达，虚拟语气，表示"如果没有……就没有办法去做某事"。

Chapter 4
Theoretical Calculation of Geotechnical Engineering Problems

岩土工程问题的理论计算

4.1 Definition of geotechnical engineering problems
岩土工程问题定义

【Text】

Deformation (engineering)

As shown in Fig. 4.1, Compressive stress results in deformation which shortens the object but also expands it outwards.

In materials science, deformation is a change in the shape or size of an object due to

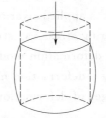

Fig. 4.1 Compressive deformation
图 4.1 受压变形

- an applied force (the deformation energy in this case is transferred through work) or
- a change in temperature (the deformation energy in this case is transferred through heat)

The first case can be a result of tensile (pulling) forces, compressive (pushing) forces, shear, bending or torsion (twisting).

In the second case, the most significant factor, which is determined by the temperature, is the mobility of the structural defects such as grain boundaries, point vacancies, line and screw dislocations, stacking faults and twins in both crystalline and non-crystalline solids. The movement or displacement of such mobile defects is thermally activated, and thus limited by the rate of atomic diffusion.

Deformation is often described as strain. As deformation occurs, internal inter-molecular forces arise that oppose the applied force. If the applied force is not too large these forces may be sufficient to completely resist the applied force, allowing the object to assume a new equilibrium state and to return to its original state when the load is removed. A larger applied force may lead to a permanent deformation of the object or even to itsstructural failure.

Fig. 4.1 describes compression deformation. Compressive stress results in deformation which shortens the object but also expands it outwards. In the figure it can be seen that the compressive loading (indicated by the arrow) has caused deformation in the cylinder so that the original shape (dashed lines) has changed (deformed) into one with bulging sides. The sides bulge because the material, although strong enough to not crack or otherwise fail, is not strong enough to support the load without change, thus the material is forced out laterally. Internal forces (in this case at right angles to the deformation) resist the applied load. The concept of a rigid body can be applied if the deformation is negligible.

Types of deformation

Depending on the type of material, size and geometry of the object, and the forces applied, various types of deformation may result. Fig. 4.2 shows the engineering stress vs. strain diagram for a typical ductile material such as steel. Different deformation modes may occur under different conditions, as can be depicted using a deformation mechanism map. Typical stress-strain diagram with the various stages of deformation is shown in Fig. 4.2.

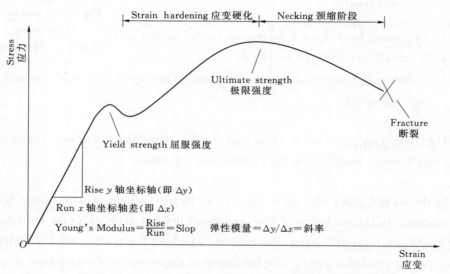

Fig. 4.2 Typical stress-strain diagram with the various stages of deformation
图 4.2 不同变形阶段的典型应力-应变关系图

1. Elastic deformation

This type of deformation is reversible. Once the forces are no longer applied, the object returns to its original shape. Elastomers and shape memory metals such as Nitinol exhibit large elastic deformation ranges, as does rubber. However elasticity is nonlinear in these materials. Normal metals, ceramics and most crystals show linear elasticity and a smaller elastic range.

Linear elastic deformation is governed by Hooke's law, which states: $\sigma = E\varepsilon$

Where σ is the applied stress, E is a material constant called Young's modulus, and ε is the resulting strain. This relationship only applies in the elastic range and indicates that the slope of the stress vs. strain curve can be used to find Young's modulus. Engineers often use this calculation in tensile tests. The elastic range ends when the material reaches its yield strength. At this point plastic deformation begins.

Note that not all elastic materials undergo linear elastic deformation; some, such as concrete, gray cast iron, and many polymers, respond in a nonlinear fashion. For these materials Hooke's law is inapplicable.

2. Plastic deformation

This type of deformation is irreversible. However, an object in the plastic deformation range will first have undergone elastic deformation, which is reversible, so the object will return part way to its original shape. Soft thermoplastics have a rather large plastic deformation range as do ductile metals such as copper, silver, and gold. Steel does, too, but not cast iron. Hard thermosetting plastics, rubber, crystals, and ceramics have minimal plastic deformation ranges. One material with a large plastic deformation range is wet chewing gum, which can be stretched dozens of times its original length.

Under tensile stress, plastic deformation is characterized by a strain hardening region and a necking region and finally, fracture (also called rupture). During strain hardening the material becomes stronger through the movement of atomic dislocations. The necking phase is indicated by a reduction in cross-sectional area of the specimen. Necking begins after the ultimate strength is reached. During necking, the material can no longer withstand the maximum stress and the strain in the specimen rapidly increases. Plastic deformation ends with the fracture of the material.

Compressive failure

Usually, compressive stress applied to bars, columns, etc. leads to shortening.

Loading a structural element or specimen will increase the compressive stress until it reaches its compressive strength. According to the properties of the material, failure modes are yielding for materials with ductile behavior (most metals, some soils and plastics) or rupturing for brittle behavior (geomaterials, cast iron, glass, etc.). Stress-strain curve of a ductile metal is shown in Fig. 4.3.

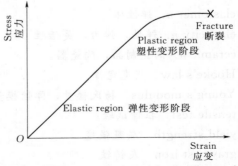

Fig. 4.3 Stress-strain curve of a ductile metal
图 4.3 延性金属应力-应变曲线图

In long, slender structural elements—such as columns or truss bars—an increase of compressive force F leads to structural failure due to buckling at lower stress than the compressive strength.

Fracture

This type of deformation is also irreversible. A break occurs after the material has reached the end of the elastic, and then plastic, deformation ranges. At this point forces accumulate until they are sufficient to cause a fracture. All materials will eventually fracture, if sufficient forces are applied.

Misconceptions

A popular misconception is that all materials that bend are "weak" and those that don't are "strong". In reality, many materials that undergo large elastic and plastic deformations, such as steel, are able to absorb stresses that would cause brittle materials, such as glass, with minimal plastic deformation ranges, to break.

【Key words】

deformation $n.$ 变形
grain boundary 晶界
point vacancy 空缺点
screw dislocation 螺旋（形）位错
stacking faults and twins 堆垛层错与孪晶
thermally $adv.$ 热地
diffusion $n.$ 扩散
equilibrium $n.$ 平衡，均势
bulge $v.$ 膨胀，凸出
at right angles to 与……成正交
ductile material 延性材料
depict $v.$ 描绘，描画
reversible $adj.$ 可逆的
elastomer $n.$ 弹性体
elasticity $n.$ 弹性，弹力，灵活性
ceramic $n.$ 陶瓷制品，陶瓷器
Hooke's law 胡克定律
Young's modulus 杨氏模量；弹性模量
tensile test 拉伸试验
yield strength 屈服强度
gray cast iron 灰铸铁
polymer $n.$ 多聚物，[高分子]聚合物
thermoplastic $n.$ 热塑性塑料

fracture n. 破裂，断裂
rupture n. 断裂，破裂
brittle adj. 易碎的
buckle vt. & vi. 用搭扣扣紧；(使)变形，弯曲

【Translation】

Deformation (engineering)　变形（工程）

As shown in Fig. 4.1, Compressive stress results in deformation which shortens the object but also expands it outwards.

如图4.1所示，压应力导致了变形，它使物件变短，但同时也使它向外扩张。

In materials science, deformation is a change in the shape or size of an object due to

在材料科学中，变形是指一个物体的形状或尺寸由于以下原因发生变化：

- an applied force (the deformation energy in this case is transferred through work)
  受力作用（在这种情况下，变形能通过做功转换）
- a change in temperature (the deformation energy in this case is transferred through heat)
  温度变化（在这种情况下，变形能通过热量转换）

The first case can be a result of tensile (pulling) forces, compressive (pushing) forces, shear, bending or torsion (twisting).

第一种情况可以是张力（拉力）、压力（推力）、剪切、弯曲或扭转（扭曲）造成的结果。

In the second case, the most significant factor, which is determined by the temperature, is the mobility of the structural defects such as grain boundaries, point vacancies, line and screw dislocations, stacking faults and twins in both crystalline and non-crystalline solids. The movement or displacement of such mobile defects is thermally activated, and thus limited by the rate of atomic diffusion.

在第二种情况下，由温度决定的最显著的因素是结构薄弱环节的流动性，比如颗粒边界、点状空缺、边线和螺丝错位、叠层间缝以及在晶体和非晶体固体中都有的孪晶。这些可动薄弱环节的运动或位移是因热激活的，从而受原子扩散速率的限制。

Deformation is often described as strain. As deformation occurs, internal inter-molecular forces arise that oppose the applied force. If the applied force is not too large these forces may be sufficient to completely resist the applied force, allowing the object to assume a new equilibrium state and to return to its original state when the load is removed. A larger applied force may lead to a permanent deformation of the object or even to its structural failure.

变形常被描述为应变。当出现变形时，内部分子间出现力，用以抵抗所施加的力。如果所施加的力不太大，这些分子力可能足以完全抵抗所受的力，可以认为物体达到一个新的平衡状态，而当荷载去除时物体能够恢复其原始状态。一个较大的作用力可能导致物体永久性变形甚至发生结构破坏。

Fig 4.1 describes compression deformation. Compressive stress results in deformation which shortens the object but also expands it outwards. In the figure it can be seen that the compressive loading (indicated by the arrow) has caused deformation in the cylinder so that the original shape (dashed lines) has changed (deformed) into one with bulging sides. The sides bulge because the material, although strong enough to not crack or otherwise fail, **is not strong enough to support** the load without change, thus the material **is forced out laterally**. Internal forces (in this case at right angles to the deformation) resist the applied load. The concept of a rigid body can be applied if the deformation is negligible.

图 4.1 描述物体受压变形，压应力导致了变形，它使物件变短，但同时也使它向外扩张。在图中可以看到，压缩载荷（箭头所示）已经引起圆柱变形，原来的形状（虚线）已经变成（变形）一个两侧外凸的物体。两侧外凸是因为材料虽然强到足以不致破裂甚至破坏，但还没有强到能纹丝不动地承受所施加的荷载，因此，材料被迫向侧向扩张。内力（在这种情况下，它与变形方向垂直）抵抗所施加的载荷。如果变形是可以忽略的，则可以运用刚体的概念。

Types of deformation 变形的类型

Depending on the type of material, size and geometry of the object, and the forces applied, various types of deformation may result. The image to the right shows the engineering stress vs. strain diagram for a typical ductile material such as steel. Different deformation modes may occur under different conditions, as can be depicted using a deformation mechanism map. Typical stress-strain diagram with the various stages of deformation is shown in Fig. 4.2.

与物体的材料种类、尺寸和几何形状以及所施加的力相关，可能出现各种不同类型的变形。图 4.2 给出了一种典型的可延展性材料，如钢材的工程应力-应变关系图。不同的条件下可能出现不同的变形模式，它可以用变形原理图来阐述。不同变形阶段的典型应力与应变关系图如图 4.2 所示。

1. Elastic deformation　弹性变形

This type of deformation is reversible. Once the forces are no longer applied, the object returns to its original shape. Elastomers and shape memory metals such as Nitinol exhibit large elastic deformation ranges, as does rubber. However elasticity is nonlinear in these materials. Normal metals, ceramics and most crystals show linear elasticity and a smaller elastic range.

这种类型的变形是可逆的。一旦作用力不再作用于物体，物体就可恢复到其原始形状。弹性纤维和形状记忆金属如镍钛合金都具有很大的弹性变形范围，跟橡胶一样。然而，这些材料表现出的弹性是非线性的。普通金属、陶瓷、大多数晶体呈线弹性并具有较小的弹性范围。

Linear elastic deformation is governed by Hooke's law, which states: $\sigma = E\varepsilon$

线弹性变形服从胡克定律，即：$\sigma = E\varepsilon$

Where σ is the appliedstress, E is a material constant called Young's modulus, and ε is the resulting strain. This relationship only applies in the elastic range and indicates that the slope of the stress vs. strain curve can be used to find Young's modulus. Engineers often use this calculation in tensile tests. The elastic range ends when the material reaches its yield strength. At this point plastic deformation begins.

式中，σ 是施加的应力，E 是材料常数，称为弹性模量，而 ε 是所产生的应变。这一关系只适用于弹性范围内，关系式表明应力-应变曲线的斜率可以用来确定弹性模量。工程师通常在拉伸试验中使用这种计算方法。当材料达到屈服强度时，弹性范围结束。从这一点开始出现塑性变形。

Note that not all elastic materials undergo linear elastic deformation; some, such as concrete, gray cast iron, and many polymers, respond in a nonlinear fashion. For these materials Hooke's law is inapplicable.

请注意，并非所有的弹性材料都会经历线弹性变形；一些材料，比如混凝土、灰铸铁和许多聚合物等是以非线性的方式响应的。对这些材料来说，胡克定律是不适用的。

2. Plastic deformation　塑性变形

This type of deformation is irreversible. However, an object in the plastic deformation range will first have undergone elastic deformation, which is reversible, so the object will return part way to its original shape. Soft thermoplastics have a rather large plastic deformation range as do ductile metals such as copper, silver, and gold. Steel does, too, but not cast iron. Hard thermosetting plastics, rubber, crystals, and ceramics have minimal plastic deformation ranges. One material with a large plastic deformation range is wet chewing gum, which can be stretched dozens of times its original length.

这种类型的变形是不可逆的。但是,一个处在塑性变形范围内的物体,首先已经经历了弹性变形,而弹性变形是可逆的,所以该物体将部分恢复到其原始形状。柔软的热塑性塑料具有相当大的塑性变形范围,就像可延性金属一样,比如铜、银、金。钢铁也是这样,但铸铁不是。硬质热成形塑料、橡胶、晶体和陶瓷具有最小的塑性变形范围。一种具有较大塑性变形范围的材料是湿的口香糖,它可以拉伸至原始长度的几十倍。

Under tensile stress, plastic deformation is characterized by a strain hardening region and a necking region and finally, fracture (also called rupture). During strain hardening the material becomes stronger through the movement of atomic dislocations. The necking phase is indicated by a reduction in cross-sectional area of the specimen. Necking begins after the ultimate strength is reached. During necking, the material can no longer withstand the maximum stress and the strain in the specimen rapidly increases. Plastic deformation ends with the fracture of the material.

在拉应力作用下,塑性变形表现为一个应变硬化阶段和一个颈缩阶段,以及最后的断裂(又称破裂)阶段。在应变硬化过程中,材料通过原子移位运动而变得更强。颈缩阶段可以由试样横截面积的减少来表明。在达到极限强度后颈缩阶段开始。在颈缩阶段,材料不再能承受最大应力,而且试样的应变迅速增加。塑性变形随着材料的断裂终结。

Compressive failure 压缩破坏

Usually, compressive stress applied to bars, columns, etc. leads to shortening.

通常情况下,作用于杆件、柱体等的压缩应力会导致它们缩短。

Loading a structural element or specimen will increase the compressive stress until it reaches its compressive strength. According to the properties of the material, failure modes are yielding for materials with ductile behavior (most metals, some soils and plastics) or rupturing for brittle behavior (geomaterials, cast iron, glass, etc.). Stress-strain curve of a ductile metal is shown in Fig. 4.3.

对一个结构单元或试样加载，在达到抗压强度之前，压应力将增加。根据材料的性质，延展性材料（大部分金属、一些土壤和塑料）的破坏模式为屈服破坏，而脆性材料（岩土材料、铸铁、玻璃等）则为断裂。延性金属的应力-应变曲线图如图4.3所示。

In long, slender structural elements—such as columns or truss bars—an increase of compressive force F leads to structural failure due to buckling at lower stress than the compressive strength.

在细长的结构单元中，比如柱或桁架杆，压力 F 的增加会导致结构破坏，因为在低于压缩强度的应力时结构就会发生屈曲。

Fracture　断裂

This type of deformation is also irreversible. A break occurs after the material has reached the end of the elastic, and then plastic, deformation ranges. At this point forces accumulate until they are sufficient to cause a fracture. All materials will eventually fracture, if sufficient forces are applied.

这种类型的变形也是不可逆的。在材料已经达到其弹性然后是塑性变形的末段时，会出现一个停顿。这时，力会积累到足以导致材料断裂。如果施加的力足够，所有材料都有可能断裂。

Misconceptions　误解

A popular misconception is that all materials that bend are "weak" and those that don't are "strong". In reality, many materials that undergo large elastic and plastic deformations, such as steel, are able to absorb stresses that would cause brittle materials, such as glass, with minimal plastic deformation ranges, to break.

一个普遍的误解是，所有能弯曲的材料都是"弱"的，而那些不会弯曲的就是"强"的。事实上，许多承受了大的弹塑性变形的材料，如钢材，能够吸收可能引起具有极小塑性变形范围的脆性材料，如玻璃，发生破裂的应力。

【Important sentences】

1. In the second case, the most significant factor, which is determined by the temperature, is the mobility of the structural defects such as grain boundaries, point vacancies, line and screw dislocations, stacking faults and twins in both crystalline and non-crystalline solids.
the most significant factor 最显著的因素，为了对 factor 进行补充和解释，后接一个 which 引导的非限制性定语从句。
2. The sides bulge because the material, although strong enough to not crack or

otherwise fail, is not strong enough to support the load without change, thus the material is forced out laterally.

be not strong enough to support… 没有强到足以承受……; be forced out laterally 被迫向侧向扩张。

4.2　Theories to resolve geotechnical engineering problems 岩土工程问题求解理论

【Text】

Uniaxial behavior

For many materials, it is convenient to study the behavior on an uniaxial sample such as a bar. On this sample, we have only one component of stress (σ), which gives rise to a deformation of the bar. From the deformation, it is possible to calculate the strain (ε).

Stress-strain relations: Idealized stress-strain response.

(a) Elastic;
(b) Elastic/ideally plastic.

For a linearly elastic material, stress is proportional to strain as shown in Fig. 4.4 (a). If the stress is removed, the bar returns to its original length and the strain becomes zero. Thus on the stress-strain diagram of Fig. 4.4 (a), the arrows indicate the loading and unloading behavior.

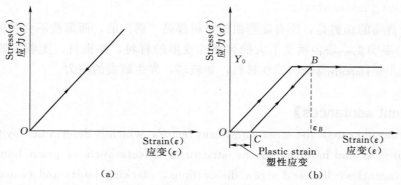

Fig. 4.4　Stress-strain relationship of elastic and ideally plastic material
图 4.4　弹性和理想塑性材料的应力-应变关系

Obviously, elastic behavior of engineering materials is observed only under a limited range of stresses and strains.

In terms of Young's modulus and Poisson's ratio, Hooke's law for isotropic materials can then be expressed as

$$\left.\begin{array}{l}\varepsilon_{11}=\dfrac{1}{E}[\sigma_{11}-\nu(\sigma_{22}+\sigma_{33})]\\[4pt]\varepsilon_{22}=\dfrac{1}{E}[\sigma_{22}-\nu(\sigma_{11}+\sigma_{33})]\\[4pt]\varepsilon_{33}=\dfrac{1}{E}[\sigma_{33}-\nu(\sigma_{11}+\sigma_{22})]\\[4pt]\varepsilon_{12}=\dfrac{1}{2G}\sigma_{12};\ \varepsilon_{13}=\dfrac{1}{2G}\sigma_{13};\ \varepsilon_{23}=\dfrac{1}{2G}\sigma_{23}\end{array}\right\} \quad (4.1)$$

In matrix form, Hooke's law for isotropic materials can be written as

$$\begin{bmatrix}\varepsilon_{11}\\ \varepsilon_{22}\\ \varepsilon_{33}\\ 2\varepsilon_{23}\\ 2\varepsilon_{13}\\ 2\varepsilon_{12}\end{bmatrix}=\begin{bmatrix}\varepsilon_{11}\\ \varepsilon_{22}\\ \varepsilon_{33}\\ \gamma_{23}\\ \gamma_{13}\\ \gamma_{12}\end{bmatrix}=\frac{1}{E}\begin{bmatrix}1 & -\nu & -\nu & 0 & 0 & 0\\ -\nu & 1 & -\nu & 0 & 0 & 0\\ -\nu & -\nu & 1 & 0 & 0 & 0\\ 0 & 0 & 0 & 2(1+\nu) & 0 & 0\\ 0 & 0 & 0 & 0 & 2(1+\nu) & 0\\ 0 & 0 & 0 & 0 & 0 & 2(1+\nu)\end{bmatrix}\begin{bmatrix}\sigma_{11}\\ \sigma_{22}\\ \sigma_{33}\\ \sigma_{23}\\ \sigma_{13}\\ \sigma_{12}\end{bmatrix} \quad (4.2)$$

Fig. 4.4 (b) shows the stress and strain relationship of an elastic/ideally plastic material. The behavior of the material is elastic up to a stress, Y_0, known as the "yield stress". At the stress Y_0, the strain in the bar become un-bounded, i.e. the bar deforms definitely and thus fails. The stress Y_0, is therefore "failure stress".

About plastic strain

For an elastic/ideally plastic material, stress can not exceed Y_0 in a stress controlled test. However, in a strain controlled test, any strain can be specified. For example, if a strain ε_B is imposed on material, the corresponding stress will still be Y_0. Suppose, if we reduce the strains, the stress follows the path indicated by BC. At the point C, the stress is 0, but there is a strain. This strain is permanent in the sense that even if we leave the bar free (without any stress), the bar is in a deformed state and has "plastic" strains.

Rheological analogues are commonly used in indicating the behaviour of materials. These are formed by mechanical components such as "spring" and "sliders". The rheological model of elasto-plastic behaviour is represented by a spring and a slider in series as shown in Fig. 4.5. The spring represents elastic behavior. The slider represents the plastic behaviour as it does not move unless the stress on it exceeds a threshold limit. Furthermore, the slider does not return to its original position if the stress is removed.

The rheological model also suggests that total strains ε in the material are composed of 2 parts: elastic strains ε_e in the spring and plastic strains ε_p in the slider. Elastic strains ε_e in the spring are recoverable on unloading; plastic strains ε_p in the slider are not recoverable on unloading. Total strains ε in the material:

Fig. 4.5 Rheological analogue of an elasto-plastic material

图 4.5 弹塑性材料的流变模型

$$\varepsilon = \varepsilon_e + \varepsilon_p \qquad (4.3)$$

which represents scalar quantities—strains in the uniaxial direction.

Eq. (4.3) can be generalized by assuming that it holds good for all components of strain in a multiaxial situation, i.e.

$$\varepsilon = [\varepsilon_x, \ \varepsilon_y, \ \varepsilon_z, \ \gamma_{xy}, \ \gamma_{yz}, \ \gamma_{zx}]^T \qquad (4.4)$$

Eq. (4.4) represents the vector of strain components; T represents a transpose and superscripts e and p denote elastic and plastic comments respectively.

Thus Eq. (4.3) represents 6 equations. Addition of elastic and plastic parts give the "total" for each of the six components of strains.

Basic ingredients of the theory of plasticity

There are four basic ingredients of the theory of plasticity:

- Stress-strain relationship prior to yielding
- Yield criterion
- Flow rule
- Hardening rule

(1) Stress-strain relationship prior to yielding

It is generally assumed that the behaviour is linear elastic prior to yielding (but it is not essential and alternative assumptions such as nonlinear elasticity can as well be made).

(2) Yield criterion

Yield criterion is a generalization of the yield point in the uniaxial case. It is a scalar function of stresses, i. e.

$$F(\sigma) = 0 \qquad (4.5)$$

where F represents a function and σ is the vector of stresses.

The components of the stress vector will change with a change of coordinate axes to say σ^*. But the yielding of material should not depend on the choice of the axes of reference if the material is isotropic, i. e.

$$F(\sigma) = F(\sigma*) = 0 \qquad (4.6)$$

This is possible only if F is a function of stress invariants and not stress components. Principal stresses (σ_1, σ_2, σ_3) are a set of stress invariants of the stress tensor. Thus, for isotropic material we have:

$$F(\sigma_1, \sigma_2, \sigma_3) = 0 \qquad (4.7)$$

A yield function represented by equation above can be plotted in the stress space as a surface which is known as "yield surface".

(3) Flow rule

In the uniaxial example, it was assumed that the plastic strain increments took place in the direction of the stress applied. In a general multidimensional situation, all components of plastic strain will exist.

(4) Hardening rule

The hardening rule specifies as to how the yield function changes with the accumulation of plastic strains. If it does not change at all we have ideally elasto-plastic material or perfectly plastic material. If the yield function changes with plastic strains, then it should be a function of stresses as well as plastic strains, i. e.

$$F(\sigma, \varepsilon_p) = 0 \tag{4.8}$$

Note:

In the multiaxial case, the yield surface moves from its original position to a new position corresponding to a higher stress state. As shown in Fig. 4.6, after yielding at a stress of Y_0, the stress and strain increase. Finally at a stress of Y_f, the strains increase indefinitely. If the material is unloaded at any point (say D) in the strain hardening region, permanent strain denoted by OE takes place. On reloading, the behavior is elastic along ED as if the yield point had shifted to point D (stress Y) due to plastic strain OE.

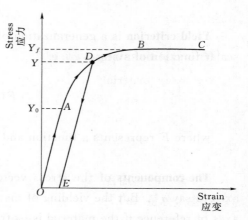

Fig. 4.6 Stress-strain relationship in the multiaxial case

图 4.6 多轴情况下的应力-应变关系

Theory of elasto-viscoplasticity

Theory of elasto-plasticity discussed in the last section does not have a "time" element. In the other words, the plastic strains are generated instantaneously. In the theory of elasto-viscoplasticity, plastic strains, called viscoplastic strains, are assumed to accrue with time. This is quite an appealing aspect of theory and is in line with experience and practice in rock mechanics. After all, monitoring of rock structures by instrumentation is based on the assumption that displacements, strains and stresses will tend to a steady state with time. It is not surprising that theory has been extensively used in the analysis of jointed rock masses.

【Key words】

Poisson's ratio　泊松比
isotropic material　各向同性材料
rheological analogue　流变模型
elasto-plastic material　弹塑性材料
threshold *adj*. 阈值的，临界值的
scalar *n*. 数量，标量
generalize *vt*. & *vi*. 概括，归纳；推广，普及
multiaxial *adj*. 有多个轴的，多轴的
transpose *vt*. [数] 移项，转置
superscript *n*. 上角文字，上标
ingredient *n*. （混合物的）组成部分；（构成）要素；因素

increment n. 增长，增量，增额
elasto-viscoplasticity 弹黏塑性
viscoplastic adj. 黏塑性的

【Translation】

Uniaxial behavior 单轴特性

For many materials, it is convenient to study the behavior on an uniaxial sample such as a bar. On this sample, we have only one component of stress (σ), which gives rise to a deformation of the bar. From the deformation, it is possible to calculate the strain (ε).

对于许多材料而言，研究其像杆一样的单轴试样是很方便的。在本示例中，只有一个应力分量（σ）引起杆件产生变形。通过变形就能够计算出应变（ε）。

Stress-strain relations: Idealized stress-strain response.

应力-应变关系：理想化的应力-应变响应。

(a) Elastic 弹性
(b) Elastic/ideally plastic 弹性/理想塑性

For a linearly elastic material, stress is proportional to strain as shown in Fig. 4.4 (a). If the stress is removed, the bar returns to its original length and the strain becomes zero. Thus on the stress-strain diagram of Fig. 4.4 (a), the arrows indicate the loading and unloading behavior.

对于线弹性材料而言，如图 4.4（a）所示，应力与应变成正比。如果去除应力，则杆件也会恢复到它的原始长度，且应变变回零。因此，在图 4.4（a）的应力-应变关系图中，箭头表明加载和卸载的特性。

Obviously, elastic behavior of engineering materials is observed only under a limited range of stresses and strains.

显然，仅在一定的应力和应变范围内能观察到工程材料的弹性特性。

In terms of Young's modulus and Poisson's ratio, Hooke's law for isotropic materials can then be expressed as

引入弹性模量和泊松比，各向同性材料的胡克定律可以表示为

$$\left.\begin{aligned}\varepsilon_{11} &= \frac{1}{E}[\sigma_{11} - \nu(\sigma_{22} + \sigma_{33})] \\ \varepsilon_{22} &= \frac{1}{E}[\sigma_{22} - \nu(\sigma_{11} + \sigma_{33})] \\ \varepsilon_{33} &= \frac{1}{E}[\sigma_{33} - \nu(\sigma_{11} + \sigma_{22})] \\ \varepsilon_{12} &= \frac{1}{2G}\sigma_{12};\ \varepsilon_{13} = \frac{1}{2G}\sigma_{13};\ \varepsilon_{23} = \frac{1}{2G}\sigma_{23}\end{aligned}\right\} \quad (4.1)$$

In matrix form, Hooke's law for isotropic materials can be written as

各向同性材料的胡克定律可以写成下面的矩阵形式

$$\begin{bmatrix}\varepsilon_{11}\\\varepsilon_{22}\\\varepsilon_{33}\\2\varepsilon_{23}\\2\varepsilon_{13}\\2\varepsilon_{12}\end{bmatrix} = \begin{bmatrix}\varepsilon_{11}\\\varepsilon_{22}\\\varepsilon_{33}\\\gamma_{23}\\\gamma_{13}\\\gamma_{12}\end{bmatrix} = \frac{1}{E}\begin{bmatrix}1 & -\nu & -\nu & 0 & 0 & 0\\-\nu & 1 & -\nu & 0 & 0 & 0\\-\nu & -\nu & 1 & 0 & 0 & 0\\0 & 0 & 0 & 2(1+\nu) & 0 & 0\\0 & 0 & 0 & 0 & 2(1+\nu) & 0\\0 & 0 & 0 & 0 & 0 & 2(1+\nu)\end{bmatrix}\begin{bmatrix}\sigma_{11}\\\sigma_{22}\\\sigma_{33}\\\sigma_{23}\\\sigma_{13}\\\sigma_{12}\end{bmatrix} \quad (4.2)$$

Fig. 4.4 (b) shows the stress and strain relationship of an elastic/ideally plastic material. The behavior of the material is elastic up to a stress, Y_0, known as the "yield stress". At the stress Y_0, the strain in the bar become un-bounded, i.e. the bar deforms definitely and thus fails. The stress Y_0, is therefore "failure stress".

图 4.4（b）表示一种弹性/理想塑性材料的应力和应变关系。直至一个称为 "屈服应力" 的应力 Y_0，材料的特性都是弹性的。在应力 Y_0 处，杆件的应变不受约束，即，杆件可以一直变形直至破坏。应力 Y_0 因此也被称为 "破坏应力"。

About plastic strain 关于塑性应变

For an elastic/ideally plastic material, stress can not exceed Y_0 in a stress controlled test. However, in a strain controlled test, any strain can be specified. For example, if a strain ε_B is imposed on material, the corresponding stress will still be Y_0. Suppose, if we reduce the strains, the stress follows the path indicated by BC. At the point C, the stress is 0, but there is a strain. This strain is permanent in the sense that **even if** we leave the bar free (without any stress), the bar is in a deformed state and has "plastic" strains.

对于弹性/理想塑性材料而言，在应力控制试验中应力不能超过 Y_0。然而，在应变控制试验中，可以指定任何应变值。例如，如果一个应变 ε_B 施加于材料，其

相应的应力仍将是 Y_0。假设，如果我们减少应变，应力将沿着 BC 所指出路径返回。在 C 点，应力为 0，但仍有应变。该应变是永久性的，意味着即使杆件处于无约束状态（没有任何应力），杆件仍处在已变形状态，并有"塑性"应变。

Rheological analogues are commonly used in indicating the behaviour of materials. These are formed by mechanical components such as "spring" and "sliders". The rheological model of elasto-plastic behaviour is represented by a spring and a slider in series as shown in Fig. 4.5. The spring represents elastic behavior. The slider represents the plastic behaviour as it does not move **unless** the stress on it exceeds a threshold limit. Furthermore, the slider does not return to its original position if the stress is removed.

各种流变模型通常用来表示材料的特性。这些流变模型都是由力学元件，如"弹簧"和"滑块"所组成。如图 4.5 所示，弹塑性特性的流变模型可由一个弹簧加一个滑块串联组成。弹簧代表弹性特性。滑块代表塑性特性，因为它不会滑动，除非作用在它上面的应力超过某个触发界值。此外，如果应力被去除，滑块也不会恢复到它的初始位置。

The rheological model also suggests that total strains ε in the material are composed of 2 parts: elastic strains ε_e in the spring and plastic strains ε_p in the slider. Elastic strains ε_e in the spring are recoverable on unloading; plastic strains ε_p in the slider are not recoverable on unloading. Total strains ε in the material:

流变模型也认为，材料中总应变 ε 是由下述 2 部分组成的：弹簧上的弹性应变 ε_e 和在滑块上的塑性应变 ε_p。在弹簧上的弹性应变 ε_e 在卸载时是可恢复的；在滑块上的塑性应变 ε_p 在卸载时是不可恢复的。材料的总应变 ε 是：

$$\varepsilon = \varepsilon_e + \varepsilon_p \tag{4.3}$$

which represents scalar quantities—strains in the uniaxial direction.

上式为标量——单轴方向上的应变值。

Eq. (4.3) can be generalized by assuming that it holds good for all components of strain in a multiaxial situation, i.e.

对式（4.3）能够通过假定它适合多坐标轴情况下所有的应变分量而加以推广，即

$$\varepsilon = [\varepsilon_x, \varepsilon_y, \varepsilon_z, \gamma_{xy}, \gamma_{yz}, \gamma_{zx}]^T \tag{4.4}$$

Eq. (4.4) represents the vector of strain components; T represents a transpose and superscripts e and p denote elastic and plastic comments respectively.

式（4.4）中为应变分量向量。T 表示转置，而下标 e 和 p 分别表示弹性和塑性注释。

Thus Eq. (4.3) represents 6 equations. Addition of elastic and plastic parts give the "total" for each of the six components of strains.

因此，式（4.3）代表 6 个等式。对于 6 个应变分量的每一个来说，都是用弹性部分加塑性部分求出"总和"。

Basic ingredients of the theory of plasticity 塑性理论的基本内容

There are four basic ingredients of the theory of plasticity:

塑性理论有四项基本内容：

- Stress-strain relationship prior to yielding
  屈服前应力-应变的关系
- Yield criterion
  屈服准则
- Flow rule
  流动律
- Hardening rule
  硬化律

(1) Stress-strain relationship prior to yielding 屈服前应力-应变的关系

It is generally assumed that the behaviour is linear elastic prior to yielding (but it is not essential and alternative assumptions such as nonlinear elasticity can as well be made).

一般假定屈服之前材料具有线弹性特性（但这不是必须的，也可采用非线弹性之类的其他假定）。

(2) Yield criterion 屈服准则

Yield criterion is a generalization of the yield point in the uniaxial case. It is a scalar function of stresses, i. e.

屈服准则是单轴情况下屈服点的推广。这是应力的标量函数，即

4.2 Theories to resolve geotechnical engineering problems 岩土工程问题求解理论

$$F(\sigma) = 0 \tag{4.5}$$

where F represents a function and σ is the vector of stresses.

其中 F 代表函数，而 σ 是应力矢量。

The components of the stress vector will change with a change of coordinate axes to say σ^*. But the yielding of material should not depend on the choice of the axes of reference if the material is isotropic, i.e.

应力矢量的元素，比如 σ^*，会随着坐标轴的变化而变化。但是，如果材料是各向同性的，其屈服不应取决于所选择的参考坐标轴，即

$$F(\sigma) = F(\sigma^*) = 0 \tag{4.6}$$

This is possible only if F is a function of stress invariants and not stress components. Principal stresses (σ_1, σ_2, σ_3) are a set of stress invariants of the stress tensor. Thus, for isotropic material we have：

上式只有当 F 是一个应力不变量（与坐标无关）的函数而不是应力分量的函数时，才有可能成立。主应力（σ_1, σ_2, σ_3）是应力张量的一组应力不变量。因此，对于各向同性的材料，应有：

$$F(\sigma_1, \sigma_2, \sigma_3) = 0 \tag{4.7}$$

A yield function represented by equation above can be plotted in the stress space as a surface which is known as "yield surface".

上式代表的屈服方程在应力空间里能够绘制成一张被称为"屈服面"的表面。

(3) Flow rule 流动律

In the uniaxial example, it was assumed that the plastic strain increments took place in the direction of the stress applied. In a general multidimensional situation, all components of plastic strain will exist.

在单轴情况下，假定塑性应变增量发生在所施加的应力的方向。在一般的多维情况下，塑性应变的所有分量都会存在。

(4) Hardening rule 硬化律

The hardening rule specifies as to how the yield function changes with the accumulation of plastic strains. If it does not change at all we have ideally elasto-plastic material or perfectly plastic material. If the yield function changes with plastic strains, then it should be a function of stresses as well as plastic strains, i. e.

硬化律详细说明了屈服函数如何随塑性应变的积累而变化。如果屈服函数完全不发生变化，则是理想的弹塑性材料或完全塑性材料。如果屈服函数随着塑性应变而变化，那么它跟塑性应变一样应该是一个应力函数，即

$$F(\sigma, \varepsilon_p) = 0 \tag{4.8}$$

Note 请注意：

In the multiaxial case, the yield surface moves from its original position to a new position corresponding to a higher stress state. As shown in Fig. 4.6, after yielding at a stress of Y_0, the stress and strain increase. Finally at a stress of Y_f, the strains increase indefinitely. If the material is unloaded at any point (say D) in the strain hardening region, permanent strain denoted by OE takes place. On reloading, the behavior is elastic along ED as if the yield point had shifted to point D (stress Y) due to plastic strain OE.

在多轴的情况下，屈服面会从原来的位置移动到与一种较高应力状态对应的新位置。如图 4.6 所示，在达到应力 Y_0 值发生屈服后，应力和应变增加。最后在应力 Y_f 处，应变出现无限增加。如果材料在应变硬化区的任意位置（比如 D 点）卸载，会出现由 OE 段代表的永久变形。在重新加载时，ED 段代表弹性特性，而屈服点由于塑性应变 OE 段的影响已经改变到应力 Y 对应的 D 点。

Theory of elasto-viscoplasticity 弹黏塑性理论

Theory of elasto-plasticity discussed in the last section does not have a "time" element. In the other words, the plastic strains are generated instantaneously. In the theory of elasto-viscoplasticity, plastic strains, called viscoplastic strains, are assumed to accrue with time. This is quite an appealing aspect of theory and is in line with experience and practice in rock mechanics. After all, monitoring of rock structures by instrumentation is based on the assumption that displacements, strains and stresses will tend to a steady state with time. It is not surprising that theory has been extensively used in the analysis of jointed rock masses.

前面讨论的弹塑性理论没有考虑"时间"因素。换句话说，塑性应变是瞬间产

生的。在弹黏塑性理论中，塑性应变被称为黏塑性应变，假设随着时间而增加。这一点是该理论相当有吸引力的方面，也符合岩石力学的经验和实践。毕竟，所有用仪器对岩石结构进行的监测都是基于位移、应变和应力，都将随着时间而趋于稳定状态的假设。该理论能在节理岩体分析中得到广泛应用就不足为奇了。

【Important sentences】

1. This strain is permanent in the sense that even if we leave the bar free (without any stress), the bar is in a deformed state and has "plastic" strains.
   even if 引导让步状语从句。
2. The slider represents the plastic behavior as it does not move unless the stress on it exceeds a threshold limit.
   unless 引导条件状语从句。

4.3 Theoretical errors and geotechnical engineering needs 理论误差与岩土工程需要的关系

【Text】

Factor of safety and reliability in geotechnical engineering

Simple reliability analyses, involving neither complex theory nor unfamiliar terms, can be used in routine geotechnical engineering practice. These simple reliability analyses require little effort beyond that involved in conventional geotechnical analyses. They provide a means of evaluating the combined effects of uncertainties in the parameters involved in the calculations, and they offer a useful supplement to conventional analyses. The additional parameters needed for the reliability analyses—standard deviations of the parameters—can be evaluated using the same amount of data and types of correlations that are widely used in geotechnical engineering practice. Example applications to stability and settlement problems illustrate the simplicity and practical usefulness of the method.

The factors of safety used in conventional geotechnical practice are based on experience, which is logical. However, it is common to use the same value of factor of safety for a given type of application, such as long-term slope stability, without regard to the degree of uncertainty involved in its calculation. Through regulation or tradition, the same value of safety factor is often applied to conditions that involve widely varying degrees of uncertainty. This is not logical. Reliability calculations provide a means of evaluating the combined effects of uncertainties, and a means of distinguishing between conditions where uncertainties are particularly high or low. In spite of the fact that it has potential value, reliability theory has not been much used in routine geotechnical practice. There are two reasons for this. First, reliability theory involves terms and concepts that are not familiar to most geotechnical engineers. Second, it is commonly perceived that using reliability theory would require more data, time and effort than are available in most circumstances.

Christian et al. (1994), Tang et al. (1999) and others have described excellent examples of use of reliability in geotechnical engineering, and clear expositions of the underlying theories. The reliability concepts can be applied in simple ways, without more data, time, or effort than are commonly available in geotechnical engineering practice. Working with the same quantity and types of data, and the same types of engineering judgments that are used in conventional analyses, it is possible to make approximate but useful evaluations of reliability.

The results of simple type of reliability analyses will be neither more nor less accurate than conventional deterministic analyses that use the same types of data, judgments, and approximations. While neither deterministic nor reliability analyses are precise, they both have value and each enhances the value of the other.

It is not advocated here that factor of safety analyses be abandoned in favor of reliability analyses. Instead, it is suggested that factor of safety and reliability be used together, as complementary measures of acceptable design.

The simple types of reliability analyses require only modest extra effort as compared to that required to evaluate factors of safety, but they will add considerable value to the results of the analyses.

【Key words】
reliability n. 可靠，可信赖
combined effect 联合效应
standard deviation 标准偏差
settlement problem 沉降问题
deterministic analysis 确定性分析
approximation n. 接近，[数] 近似值

【Translation】

Factor of safety and reliability in geotechnical engineering
岩土工程安全性和可靠度

Simple reliability analyses, involving neither complex theory **nor** unfamiliar

4.3 Theoretical errors and geotechnical engineering needs 理论误差与岩土工程需要的关系

terms, can be used in routine geotechnical engineering practice. These simple reliability analyses require little effort beyond that involved in conventional geotechnical analyses. They provide a means of evaluating the combined effects of uncertainties in the parameters involved in the calculations, and they offer a useful supplement to conventional analyses. The additional parameters needed for the reliability analyses—standard deviations of the parameters—can be evaluated using the same amount of data and types of correlations that are widely used in geotechnical engineering practice. Example applications to stability and settlement problems illustrate the simplicity and practical usefulness of the method.

简单的可靠度分析，既不涉及复杂的理论，也不包含陌生的术语，可以被用在常规岩土工程实践中。这些简单的可靠度分析与传统的岩土工程分析相比并不需要付出更多努力。它们提供了一种评估计算中所涉及的参数的不确定性的综合影响的方法，而且对传统的分析提供了一种有用的补充。可靠度分析所需的附加参数——参数的标准差——可以使用岩土工程实践中广泛使用的相同量的数据和相关性类型来进行评估。稳定性和沉降问题上的实例应用说明了这一方法的简单性和实用性。

The factors of safety used in conventional geotechnical practice are based on experience, which is logical. **However, it is common to** use the same value of factor of safety for a given type of application, **such as long-term** slope stability, without regard to the degree of uncertainty involved in its calculation. Through regulation or tradition, the same value of safety factor is often applied to conditions that involve widely varying degrees of uncertainty. This is not logical. Reliability calculations provide a means of evaluating the combined effects of uncertainties, and a means of distinguishing between conditions where uncertainties are particularly high or low. In spite of the fact that it has potential value, reliability theory has not been much used in routine geotechnical practice. There are two reasons for this. First, reliability theory involves terms and concepts that are not familiar to most geotechnical engineers. Second, **it is commonly perceived that using** reliability theory would require more data, time and effort than are available **in most circumstances.**

在传统的岩土工程实践中使用的安全因素都是以经验为基础的，这也是合乎逻辑的。可是，通常的做法是对于给定类型的应用采用相同的安全系数，如长期边坡稳定性，而不考虑计算中牵涉的不确定性程度。通过规范或传统，相同的安全系数值常常用在所涉及的不确定性程度变化很大的情况，这是不合逻辑的。可靠度计算提供了一种评估不确定性的综合影响的手段，和一种区分特别高或低的不确定性的方法。尽管可靠度理论具有潜在价值，但它还没有在常规的岩土工程实践中获得大量应用。造成这一结果的原因有两个。首先，可靠度理论涉及一些对大多数岩土工程师来说不熟悉的术语和概念。其次，人们普遍以为，使用可靠度理论会需要比在大多数情况下可获得的更多的数据、时间和精力。

Christian et al. (1994), Tang et al. (1999) and others have described excellent examples of use of reliability in geotechnical engineering, and clear expositions of the underlying theories. The reliability concepts can be applied in simple ways, without more data, time, or effort than are commonly available in geotechnical engineering practice. Working with the same quantity and types of data, and the same types of engineering judgments that are used in conventional analyses, it is possible to make approximate but useful evaluations of reliability.

克里斯蒂安等（1994）、唐等（1999），还有其他人都已经介绍过将可靠度用于岩土工程领域的杰出案例，以及建立在该理论基础上的一些清晰的阐述。可靠度的概念可以用简单的方式获得应用，与一般岩土工程实践相比，并不需要更多的数据、时间或精力。在传统分析中所使用的相同数量和类型的数据以及相同类型的工程评价标准上开展工作，能够作出相近而且有用的可靠度评估。

The results of simple type of reliability analyses will be neither more nor less accurate than conventional deterministic analyses that use the same types of data, judgments, and approximations. While neither deterministic nor reliability analyses are precise, they both have value and each enhances the value of the other.

简单类型的可靠度分析结果与使用相同类型的数据、评价标准和近似假设的传统的确定性分析相比，其精度不相上下。尽管确定性分析和可靠度分析都不是精确的，但都有自身存在的价值，而且各自都能促进对方价值的提升。

It is not advocated here that factor of safety analyses be abandoned in favor of reliability analyses. Instead, it is suggested that factor of safety and reliability be used together, as complementary measures of acceptable design.

这里不提倡为了迎合可靠度分析而放弃安全性分析。相反，建议将二者结合起来，作为可行性设计的互补措施。

The simple types of reliability analyses require only modest extra effort as compared to that required to evaluate factors of safety, but they will add considerable value to the results of the analyses.

本文中所描述的简单类型的可靠性分析，只要求比进行安全系数评估付出稍多一些的努力，但他们也会大幅增加分析结果的价值。

4.3 Theoretical errors and geotechnical engineering needs   理论误差与岩土工程需要的关系

【Important sentences】

1. Simple reliability analyses, involving neither complex theory nor unfamiliar terms, can be used in routine geotechnical engineering practice.
   involving 为非谓语动词,是现在分词作状语;neither … nor,固定搭配,表示"既不……也不";句子的主干为 Simple reliability analyses can be used in routine geotechnical engineering practice。

2. However, it is common to use the same value of factor of safety for a given type of application, such as long-term slope stability, without regard to the degree of uncertainty involved in its calculation.
   However 表转折;it is common to do sth. 通常的做法是……,或者翻译为"做什么事情是很普遍的";such as 表列举,此处 such as long-term slope stability 作插入语;long-term 长期的。

3. Second, it is commonly perceived that using reliability theory would require more data, time, and effort than are available in most circumstances.
   it is commonly perceived that … 人们普遍以为……,that 引导宾语从句;using 为动名词,和 reliability theory 一起作宾语从句的主语;in most circumstances 为固定搭配,表示"在大多数情况下"。

Chapter 5
Geotechnical Engineering Design

岩土工程设计

5.1 Rules to be followed 必须遵循的准则

【Text】

The aim of structural design should be to provide, with due regard to economy, a structure capable of fulfilling its intended function and sustaining the specified loads for its intended life. The design should facilitate safe fabrication, transport, handling and erection. It should also take account of the needs of future maintenance, final demolition, recycling and reuse of material & other sustainability issues.

When defining the design situations and limit states, the following factors should be considered:

-site conditions with respect to overall stability and ground movements;

-nature and size of the structure and its elements, including any special requirements such as the design life;

-conditions with regard to its surroundings (e.g.: neighbouring structures, traffic, utilities, vegetation, hazardous chemicals);

-ground conditions;

-ground-water conditions;

-regional seismicity;

-influence of the environment (hydrology, surface water, subsidence, seasonal changes of temperature and moisture).

In geotechnical design, the detailed specifications of design situations should include, as appropriate:

-the actions, their combinations and load cases;

-the general suitability of the ground on which the structure is located with respect to overall stability and ground movements;

-the disposition and classification of the various zones of soil, rock and elements of construction, which are involved in any calculation model;

-dipping bedding planes;

-mine working, caves or other underground structures;

-in the case of structures resting on/near rock:

- interbedded hard and soft strata
- faults, joints and fissures
- possible instability of rock blocks
- solution cavities, such as swallow holes or fissures filled with soft material, and continuing solution processes

-the environment within which the design is set, including the following:

- effects of scour, erosion and excavation, leading to changes in the geometry of the ground surface
- effects of chemical corrosion
- effects of freezing
- effects of long duration droughts
- variations in ground-water levels, including, e.g. the effects of dewatering, possible flooding, failure of drainage systems, water exploitation
- the presence of gases emerging from the ground
- other effects of time and environment on the strength and other proprieties of materials; e.g. the effect of holes created by animal activities

- earthquakes;

- ground movements caused by subsidence due to mining or other activities;

- the sensitivity of the structure to deformations;

- the effect of the new structure on existing structure, services and the local environment.

【Key words】

fabrication n. 制造，捏造
erection n. 建立，设立
demolition n. 毁坏，破坏
overall stability 整体稳定性
vegetation n. 植物（总称），草木

hazardous chemical 危险药品（化学品）
seismicity n. 地震活动，受震强度
subsidence n. 沉淀，陷没，下沉
specification n. 规格，详述，说明书
interbedded adj. 层间的，夹层之间的
strata n. 地层，岩层（stratum 的复数）
fissure n. 狭长裂缝或裂隙
cavity n. 腔，洞
scour n. 擦，冲刷
dewatering n. 去水（作用），脱水（作用）
mining n. 采矿（业）
sensitivity n. 敏感，感受性，灵敏性

【Translation】

The aim of structural design should be to provide, **with due regard to economy**, a structure capable of fulfilling its intended function and sustaining the specified loads for its intended life. The design should facilitate safe fabrication, transport, handling and erection. It should also **take account of** the needs of future maintenance, final demolition, recycling & reuse of material and other sustainability issues.

结构设计的目标应该是在适当考虑经济条件的情况下，提供一个能够满足其预期的功能并在预期使用寿命期内能承受指定荷载的结构。设计应便于安全生产、运输、操作和架设。它还应考虑到未来维护、最后拆除、回收和对材料的再利用以及其他可持续性方面的问题。

When defining the design situations and limit states, the following factors should be considered:

在确定设计情况和限制状态时，应考虑以下因素：

-site conditions with respect to overall stability and ground movements;

与整体稳定性和地下位移相关的现场条件；

-nature and size of the structure and its elements, including any special requirements such as the design life;

结构及其部件的性质和尺寸，包括诸如设计寿命等的任何特殊要求；

-conditions with regard to its surroundings (e.g. neighbouring structures, traffic, utilities, vegetation, hazardous chemicals);

有关其周围环境的条件（例如：邻近结构、交通、公用设施、植被、危险化学品等）；

-ground conditions;

地下条件；

-ground-water conditions;

地下水条件；

-regional seismicity;

区域地震活动；

-influence of the environment (hydrology, surface water, subsidence, seasonal changes of temperature and moisture).

环境的影响（水文、地表水、沉降、季节性的温度和湿度的变化）。

In geotechnical design, the detailed specifications of design situations should include, as appropriate:

在岩土工程设计中，设计情况的详细说明应包括以下根据情况选择的合适项：

-the actions, their combinations and load cases;

各种作用力及其组合以及荷载条件；

-the general suitability of the ground on which the structure is located with respect to overall stability and ground movements;

与整体稳定性和地下位移相关的结构所在地层的一般适用性；

-the disposition and classification of the various zones of soil, rock and elements of construction, which are involved in any calculation model;

在任何计算模型中涉及的各区域土体、岩石和构筑单元的分布与分类；

-dipping bedding planes;

倾斜层理面；

-mine working, caves or other underground structures;

矿山开采、洞穴或其他地下结构；

-in the case of structures resting on/near rock:

在结构位于岩石上或靠近岩石的情况下：

- interbedded hard and soft strata
  软硬交错的地层
- faults, joints and fissures
  断层、节理和裂隙
- possible instability of rock blocks
  可能不稳定的岩石块
- solution cavities, such as swallow holes or fissures filled with soft material, and continuing solution processes
  溶洞，如被软质物质填充的浅的坑或裂缝，以及仍在继续的溶解进程

- the environment within which the design is set, including the following:

设计项目所处的环境，包括以下几种：

- effects of scour, erosion and excavation, leading to changes in the geometry of the ground surface
  导致地表几何形状改变的冲刷、侵蚀和开挖影响
- effects of chemical corrosion
  化学腐蚀的影响
- effects of freezing
  冻结的影响
- effects of long duration droughts
  长期干旱的影响
- variations in ground-water levels, including, e.g. the effects of dewatering, possible flooding, failure of drainage systems, water exploitation
  地下水位的变化，包括如降水、可能的洪水、排水系统的失效、水的开采利用等的影响
- the presence of gases emerging from the ground

存在自地下涌出的气体
- other effects of time and environment on the strength and other propreties of materials; e. g. the effect of holes created by animal activities
时间和环境对材料的强度和其他性能造成的其他影响；如动物活动形成的孔洞的影响

-earthquakes;

地震；

-ground movements caused by subsidence due to mining or other activities;

由于矿山开采或其他活动造成的沉降引起的地层移动；

-the sensitivity of the structure to deformations;

结构对变形的敏感性；

-the effect of the new structure on existing structure, services and the local environment.

新结构对现有结构、服务和局部环境的影响。

【Important sentences】
1. The aim of structural design should be to provide, with due regard to economy, a structure capable of fulfilling its intended function and sustaining the specified loads for its intended life.
   with due regard to economy 在适当考虑经济条件的情况下，作插入语。
2. It should also take account of the needs of future maintenance, final demolition, recycling & reuse of material and other sustainability issues.
   take account of 考虑。

5.2 General design methods 一般设计方法

【Text】

Geotechnical design by calculation

(1) Design by calculation involves:

- actions, which may be either imposed loads or imposed displacements, e. g. from ground movements
- properties of soils, rocks and other materials
- geometrical data
- limiting values of deformations, crack widths, vibrations etc.
- calculation models

(2) It should be considered that knowledge of the ground conditions depends on the extent and quality of the geotechnical investigations. Such knowledge and the control of workmanship are usually more significant to fulfilling the fundamental requirements than the precision in the calculation models and partial factors.

(3) The calculation model shall describe the assumed behaviour of the ground for the limit state under consideration.

(4) If no reliable calculation model is available for a specific limit state, analysis of another limit state shall be carried out using factors to ensure that exceeding the specific limit state considered is sufficiently improbable. Alternatively, design by prescriptive measures, experimental models and load tests, or the observational method, shall be performed.

(5) The calculation model may consist of any of the following:

- an analytical model
- a semi-empirical model
- a numerical model

(6) Any calculation model shall be either accurate or err on the side of safety.

(7) A calculation model may include simplifications.

(8) If needed, a modification of the results from the model may be used to ensure that the design calculation is either accurate or err on the side of safety.

(9) If the modification of the results makes use of a model factor, it should be take account of the following:

- the rang of uncertainty in the results of the method of analysis
- any systematic errors known to be associated with the method of analysis

(10) If an empirical relationship is used in the analysis, it shall be clearly established that it is relevant for the prevailing ground conditions.

(11) Limit states involving the formation of a mechanism in the ground should be readily checked using a calculation model.

For limit states defined by deformation considerations, the deformations should be assessed.

Note: Many calculation models are based on the assumption of a sufficiently ductile performance of the ground/ structure system.

A lack of ductility, however, will lead to an ultimate limit state characterized by sudden collapse.

(12) Numerical methods can be appropriate if compatibility of strains of the interaction between the structure and the soil at a limit state is considered.

(13) Compatibility of strains at a limit state should be considered. Detailed analysis, allowing for the relative stiffness of structure and ground, may be needed in cases where a combined failure of structural members and the ground could occur. Examples include raft foundations, laterally loaded piles and flexible retaining walls. Particular attention should be paid to strain compatibility for materials that are brittle or that have strain-softening properties.

(14) In some problems, such as excavations supported by anchored or strutted flexible walls, the magnitude and distribution of earth pressures, internal structural forces and bending moments depend to a great extent on the stiffness of the structure, the stiffness and strength of the ground and the state of stress in the ground.

(15) In these problems of ground-structure interaction, analyses should use stress-strain relationships for ground and structural materials and stress states in the ground that are sufficiently representative, for the limit state considered, to give a safe result.

Design by prescriptive measures

(1) In design situations where calculation models are not available or not necessary, exceeding limit states may be avoided by the use of prescriptive measures. These involve conventional and generally conservative rules in the design, and at-

tention to specification and control of materials workmanship, protection and maintenance procedures.

Note: Reference to such conventional and generally conservative rules may be given in the codes such as the national annex.

(2) Design by prescriptive measures may be used where comparable experience makes design calculations unnecessary. It may also be used to ensure durability against frost action and chemical or biological attack, for which direct calculations are not generally appropriate.

【Key words】

EN 1990: 2002　土木工程设计施工欧盟标准规范
geometrical *adj*. 几何的，几何学的
workmanship *n*. 技艺，工艺
semi-empirical *adj*. 半经验的
simplification *n*. 单纯化，简单化
ductility *n*. 展延性，柔软性
annex *n*. 附录，附件
comparable experience　相关专业

【Translation】

Geotechnical design by calculation 岩土设计计算

(1) Design by calculation involves:

设计计算涉及：

- actions, which may be either imposed loads or imposed displacements, e.g. from ground movements
  各种作用力，可能由施加的荷载或强制的位移引起，例如，因地层运动引起
- properties of soils, rocks and other materials
  土、岩石和其他材料的性质
- geometrical data
  几何数据
- limiting values of deformations, crack widths, vibrations etc.
  变形、裂缝宽度、振动等的极限值
- calculation models
  计算模型

(2) It should be considered that knowledge of the ground conditions depends on the extent and quality of the geotechnical investigations. Such knowledge and the control of workmanship are usually **more significant to fulfilling** the fundamental requirements **than** the precision in the calculation models and partial factors.

应该考虑到，对地层条件的了解情况取决于岩土工程勘察的程度和质量。在满足基本需求方面，对岩土工程勘察工艺的了解和掌控常常比计算模型和某些参数的精度更重要。

(3) The calculation model shall describe the assumed behaviour of the ground for the limit state under consideration.

计算模型应该能描述在所考虑的极限状况下地层的假定响应特性。

(4) **If** no reliable calculation model **is available for** a specific limit state, **analysis of another limit state shall be carried out** using factors **to ensure that** exceeding the specific limit state **considered** is sufficiently improbable. Alternatively, design by prescriptive measures, experimental models and load tests, or the observational method, shall be performed.

如果没有可靠的计算模型可用于特定的极限状态，应使用能足以保证不太可能超出所考虑的特定极限状态的系数进行另外的极限状态分析。或者，在进行设计时应该采取指定的措施、实验模型和荷载测试或观测方法。

(5) The calculation model may consist of any of the following:

计算模型可能由下述模型中的任何一个组成：

- an analytical model
  分析模型
- a semi-empirical model
  半经验模型
- a numerical model
  数值模型

(6) Any calculation model shall be either accurate or wrong on the side of safety.

站在安全角度考虑，任何计算模型应该不是准确的就是错误的。

(7) A calculation model may include simplifications.

一个计算模型可能包括一些简化。

(8) If needed, a modification of the results from the model may be used to ensure that the design calculation is either accurate or wrong on the side of safety.

如果有需要，站在安全的角度考虑，为了保证设计计算要么准确要么错误，可能要对计算模型给出的结果进行修改。

(9) If the modification of the results makes use of a model factor, it should be take account of the following:

如果结果的修正使用了某个模型参数，则应考虑以下情形：

- the rang of uncertainty in the results of the method of analysis
  分析方法所得结果中的不确定度的范围
- any systematic errors known to be associated with the method of analysis
  已知的与分析方法有关的任何系统误差

(10) If an empirical relationship is used in the analysis, it shall be clearly established that it is relevant for the prevailing ground conditions.

如果在分析中使用经验关系，则要清楚地确定它与最有代表性的地基条件是相对应的。

(11) Limit states involving the formation of a mechanism in the ground should be readily checked using a calculation model.

地基中涉及机制形成的极限状态应能方便地使用计算模型进行校验。

For limit states defined by deformation considerations, the deformations should be assessed.

对于通过考虑变形来定义的极限状态，应该对变形进行评估。

Note: Many calculation models are based on the assumption of a sufficiently ductile performance of the ground/structure system.

注意：许多计算模型建立在地基或结构系统具有足够的延展性的假定基础上。

A lack of ductility, however, will lead to an ultimate limit state characterized by sudden collapse.

然而，缺乏延展性将会导致一个以突然崩溃为特点的终极极限状态。

(12) Numerical methods can be appropriate if compatibility of strains of the interaction between the structure and the soil at a limit state is considered.

如果考虑在极限状态下结构和土壤之间相互作用的应变相容性，数值方法可能是合适的。

(13) Compatibility of strains at a limit state should be considered. Detailed analysis, allowing for the relative stiffness of structure and ground, may be needed in cases where a combined failure of structural members and the ground could occur. Examples include raft foundations, laterally loaded piles and flexible retaining walls. Particular attention should be paid to strain compatibility for materials that are brittle or that have strain-softening properties.

在极限状态下的应变相容性应该予以考虑。考虑到结构和地基的相对刚度，在结构构件和地基上有可能发生复合破坏的情况下，可能需要进行详细的分析。工程案例包括筏型基础、侧向承载桩和柔性挡土墙。对于脆性或具有应变软化特性的材料需要特别关注应变相容性。

(14) In some problems, such as excavations supported by anchored or strutted flexible walls, the magnitude and distribution of earth pressures, internal structural forces and bending moments depend to a great extent on the stiffness of the structure, the stiffness and strength of the ground and the state of stress in the ground.

对于一些问题，比如有锚固或有支柱翼的柔性挡土墙支护的开挖、土压力的大小和分布、内部结构力和弯矩，很大程度上取决于结构的刚度、地基的刚度和强度以及地基中的应力状态。

(15) In these problems of ground-structure interaction, analyses should use stress-strain relationships for ground and structural materials and stress states in the ground that are sufficiently representative, for the limit state considered, to give a safe result.

对于地基和结构相互作用的这些问题，为了给出安全的结果，针对所考虑的极限状态，分析时应采用有足够代表性的地基与结构材料间的应力-应变关系以及地基中的应力状态。

5.2 General design methods 一般设计方法 /129/

Design by prescriptive measures 按指定方法设计

(1) In design situations where calculation models are not available or not necessary, exceeding limit states may be avoided by the use of prescriptive measures. These involve conventional and generally conservative rules in the design, and attention to specification and control of materials workmanship, protection and maintenance procedures.

在计算模型不可用或不必要的设计条件下，使用指定性方法可以避免超过极限的状态。这些指定性方法涉及在设计中的一些常规的和通常偏保守的做法，以及规范的注意事项和材料的工艺控制、保护和维护流程。

Note: Reference to such conventional and generally conservative rules may be given in the codes such as the National annex.

注意：国家索引一类的文件可能提供关于这些常规的和通常偏保守的做法的参考资料。

(2) Design by prescriptive measures may be used where comparable experience makes design calculations unnecessary. It may also be used to ensure durability against frost action and chemical or biological attack, for which direct calculations are not generally appropriate.

按指定性方法设计可以用在因为有经验类比而不必进行设计计算的情况。这种方法也可以被用来确保应对霜冻作用、化学或生物侵蚀的耐久性，而直接计算法通常不胜任这些工作。

【Important sentences】

1. Such knowledge and the control of workmanship are usually more significant to fulfilling the fundamental requirements than the precision in the calculation models and partial factors.
   to fulfilling 不定式作目的状语；more significant than 比较，表更重要。
2. If no reliable calculation model is available for a specific limit state, analysis of another limit state shall be carried out using factors to ensure that exceeding the specific limit state considered is sufficiently improbable.
   be available 可以使用的；limit state 极限状态，固定短语；using 动名词作伴随状语；to ensure that，that 引导宾语从句；considered 过去分词作后置定语。

Chapter 6
Geotechnical Engineering Construction

岩土工程施工

6.1 General construction methods 一般施工方法

【Text】

Geotechnical engineering characteristics and general construction technology

Geotechnical engineering construction has the characteristics of concealment, complexity and rigour. At present, the application of geotechnical engineering construction technology mainly includes: foundation treatment, foundation engineering construction technology, slope reinforcement engineering construction technology, trenchless technology and geotechnical engineering testing technology, etc. Some of the typical construction technologies are discussed as follows.

Diaphragm walls

Diaphragm walls bear with its static function the ground and water pressure and transfer them into the ground, through anchoring or ground resistance in its lower part. First of all, the excavation pit shall be made, in which guide walls are set up. They represent a kind of template for the diaphragm wall and define construction pit shape, as shown in Fig. 6.1. Furthermore, they ensure correct direction of grapple and stabilize the upper zone of diaphragm wall. The diaphragm wall is executed in panels. In the course of diaphragm wall construction a bentonite suspension is used, which strengthens the ground through its hydraulic pressure. When advancing with the excavation works and as a consequence of suspension loss in the underlying ground, it is necessary constantly to be refilled. After the required depth is reached, reinforcement package is lowered and then the panel is concreted with concreting pump.

Fig. 6.1 Diaphragm walls
图 6.1 地下连续墙

This method treats the execution of diaphragm walls with special diaphragm wall equipment.

The diaphragm walls are formed out of reinforced concrete elements, executed one by one in the ground. This type of construction is able to perform support-, reinforcement-and water resistant functions, which determines their application field. The most frequent use of diaphragm walls is the excavation strengthening in urban conditions.

Anchoring of the ground

The fundamental concept of ground reinforcement with ground nails involves fortification of the ground through passive driving in of nails, close to each other, in order to create a coherent construction and thus increase the overall strength of ground slitting and to limit dislocation. The main idea is to transmit sustainable tensile strength, generated by tamping the ground through friction, created in the surfaces.

Self-drilling IBO anchors are successfully used for anchoring in weak soil, coherent and non-coherent soils, and in case of unstable drilling.

Self-drilling anchors are installed with drilling equipment with air or hydraulic rotation strikes, with the help of drilling wash, appropriate for specific soil conditions.

Main applications:

-Tunnel digging

- Reinforcement with rock bolts
- Anchoring of chambers
- Reinforcement in front of the tunnel chamber bolts
- Pilot with widening of the heel
- Tunnel portals, trenches and zones where the open method of construction is applied

-Special activities

- Reinforcement of excavation pits, dams and swaths with injection anchors
- Anchoring of support walls and noise barriers
- Pilot fundaments and control for their support in water—injection pilots/micropilots

Pilots/Drilling pilots

Building by means of pilot holes is one of the oldest methods. It is applied in soft or water-saturated soil or in narrow building sites. Pilots vary according to the type of building, disposition and method used.

In the process of building they are specified as single, tangent and cut-off pilots.

Reinforced packages are prepared in advance and installed in one piece or sequentially, and then they are covered with concrete in the drilled opening. Cut-off pilots are placed in the following order: reinforced—not reinforced.

Sheet piling

The sheet walls could be pressed, driven by vibrations or by hammering into the ground in compliance with the local ground characteristics. Moreover, they could be installed as dense wall, in the excavated ground diaphragm panels. Since sheet walls are executed as barriers, which are water impermeable, it is possible to meet the need of lowering ground water table on big areas through their joining with cohesive soil layer or artificial dense bottom.

Drainage

This system uses a series of wells, connected into a joint suction collector. Many single wells can be connected to one pump only (usually located in the suction collector). This type of system works effectively in variable soils and in shallow water-carrying layers, where the water level has been lowered to a waterproof layer, as shown in Fig. 6.2.

Fig. 6.2 Foundation pit drainage
图 6.2 基坑排水

Construction excavations

Construction pits are formed from excavating the soil and include basements and the underground floors when the closed method of construction is applied. This includes all elements as walls, bottoms, reinforcement and retaining water. If the underground water level is above the lowest part of the pit (excavation bottom), they have to be lowered. This is done with the help of the so-called water-lowering by means of wells. Another method is to lay concrete on the excavation bottom in the shape of a compact concrete plate made of waterproof concrete. The excavation of foundation pit is shown in Fig. 6.3.

Fig. 6.3 Foundation pit excavation
图 6.3 基坑开挖

Methods for tunneling construction

The following methods are used in tunnel construction:

(1) **Cut and cover method.** Cut and cover method in reinforced trenches and in trenches with slopes respectively, depending on the presence of buildings nearby the route, or the presence of free space for digging trenches with slopes. The construction of the metro stations or the tunnels is implemented in these trenches. After that they are backfilled and the surface above is restored, as shown in Fig. 6.4. In the so built constructions installing of the equipment and architectural outlining is done. Massive reinforcements with great extent of hardness are mainly applied, thus allowing for the construction works in close proximity to the existing buildings and facilities without compromising their integrity. In some cases these reinforcements are parts of the supporting structure of the stations and the tunnels. The first station from the section St. Nedelja Sq. to housing complex Ljulin and three of the stations from the section St. Nedelja Sq. to Mladost, were built in a similar way. The same method is planned to be used for some of the stations from the sec-

ond metro-diameter—Central Railway Station, Lavov most, National Palace of Culture and St. Naum.

Fig. 6.4 Construction of cut and cover method
图 6.4 明挖法施工

(2) **Milan method.** Milan method is applied when a rapid restoration of the surface above the metro facilities is necessary. The Milan method allows for the simultaneous building of the underground and overground parts of the facilities. This approach is described as "cover and cut/dig", which means parallel slit walls, or pilots/sheet pile walls built and then connected to a monolith plate over them, after which the soil is excavated, as shown in Fig. 6.5. The Milan method comprises both building slit walls and pilots under the base and the fundamental system, prior to launching the tunnel excavation. Concrete slit walls, built with the help of bentonite solution, are used for enclosing walls. They have a double function: side support during the construction process and construction walls of the facility. This method has been applied partly in the construction of Joliot Curie and G. M. Dimitrov stations, as well as in some parts of the tunnels close to these stations. In the second metro-diameter, in order to minimize the time for stopping or to limit the traffic on the main streets, this method is envisaged to be applied at MS-8-II in front of TSUM, and MS 11-II under Cherny vrah Blvd., in Lozenets, as well as for the facilities to change the direction of movement after it.

(3) **Shield method.** The mechanized shield method is applied in the construction of tunnels in the central part of the city, where the tunnels are of considerable length; where archaeological sites are present, as well as to avoid open excavation works of considerable length along the major boulevards. The method involves mechanized excavation of soil in the front part of a steel cylinder, called "shield", and assembling the tunnel construction under the protection of the back part of this cylinder.

/138/ Chapter 6 Geotechnical Engineering Construction 岩土工程施工

Fig. 6.5 Construction of Milan method
图 6.5 米兰法施工

The machine is pushed forward by a system of hydraulic jacks, stuck in the tunnel construction in the back part of the shield. Because of the considerable depth of tunnels, this method was applied to the section St. Nedelja Sq. -V. Levski Stadium and it is intended to be used for the tunnels in the section Road junction Nadezhda - Han Asparuh Str. before reaching Patriarch Evtimij Blvd. Due to the location of the tunnels under the level of underground waters, shields with hydraulic counter pressure in the face chamber will be used for both sections. This will prevent sinking of the ground over the tunnels and the negative effect on the building above. The shield machine, which is 85m long and 1,600t heavy, will dig 9m daily and will transport about 1,000m^3 or 50 trucks of excavated material. The structure of shield machine and excavated shield tunneling are shown in Fig. 6.6.

(a) A diagram of the structure of shied machine
盾构机结构示意图

Fig. 6.6 (1) Shield method
图 6.6 (一) 盾构法

(b) Excavated shield tunneling
开挖好的盾构隧道

Fig. 6.6 (2)　Shield method
图 6.6（二）　盾构法

(4) **New Austrian tunnelling method.** This method is applied effectively to sections with appropriate deployment of the terrain and limited water flow. It involves working on small sections of the face chamber stage by stage, then a primary construction/lining with reinforced cement solution is deployed above it and after that at certain distance a secondary construction/panelling of the tunnel is being reinforced and covered with concrete by means of a special movable formwork. The essence of the "New Austrian tunnelling method" implies transforming rock masses into support elements, as shown in Fig. 6.7. It relies on the inherent strength of the surrounding rock mass being conserved as the main component of tunnel support. The tunnel is usually dug in a full profile; however it can be excavated in parts. The main idea is through embedding deformable supportive elements—an-

Fig. 6.7　Construction of New Austrian tunneling method
图 6.7　新奥法施工

chors, steel arc frames and shotcrete—to form an alleviating arc to turn it into a support element, substituting the lining. The new stable equilibrium is determined by fading of the radial deformations of the excavation, measured on the spot. This method is applied in masses with random strength with sufficient tunnel covering to form the alleviating arc, and with tunnels with random size of the cross section. This method was applied to the link between V. Levski Stadium and the tunnel under Dragan Tsankov Blvd. in front of the National Radio. Due to the specifics of the geologic conditions, deep ground waters and significant depth of deployment of the tunnels from the second metro-diameter in the section under Cherny vrah Blvd. between Hemus Hotel and James Bourchier Blvd., the new Austrian method will be applied to their construction. In the section after the metro station under Cherny vrah Blvd., due to the significant width of the facilities, shallow deployment, high level of ground water, and the presence of ventilation system facilities above it, the application of this method is not appropriate.

The selection principles of geotechnical engineering construction technology

(1) **Economic principle**: Because of the "uncertainty" of construction technology, there are usually several sets of technology available for each type of geotechnical engineering problems. Then it needs to make comparison between the factors such as economy, time limit for a project, technology, safety etc. But in any case, the economy of technology will always dominate the selection.

(2) **Applicability principle**: There is no absolute good or bad geotechnical engineering technology, i. e. a geotechnical technology should not be judged to be absolutely good or bad only by some/several indicators of the technology. It's not necessary to use the best, but we must use the most appropriate construction technology.

(3) **Practical principle**: Because of the "uncertainty" of geotechnical engineering construction technology, the technical feasibility of construction technology must not be judged only by theoretical analysis and calculation. It's more important that the real engineering practice should be used to choose the most suitable method.

(4) **Environmental protection principle**: With the country's increasingly stringent environmental requirements, the environmental effect in the geotechnical engineering construction and the influencement degree of various construction methods to the environment will become an important reference to construction technology selection. More and more environmental protection construction technologies will be the first choice of geotechnical engineering construction.

【Key words】

static function 静态功能
ground resistance 接地电阻
guide wall 导墙
template *n*. 样板；模板；型板
grapple *n*. 扭打，格斗；紧握；抓机，抓斗
panel *n*. 镶板；面；（门、墙等上面的）嵌板
bentonite suspension 膨润土悬浮液
suspension loss 悬浮液损失
reinforcement package 加固包
concreting pump 混凝土泵
ground nail 地钉
ground slitting 地面开槽
self-drilling IBO anchor IBO 钻锚杆
tunnel chamber 隧道硐室
swath *n*. 收割的刈痕，细长的列；草条；刈幅；割幅
injection anchor 喷锚
pilot fundament 试点基础
avalanche construction 塌方构造
pilot hole 导孔
ground water table 地下水位，潜水位
suction collector 吸收器
water-carrying layer 载水层
retaining water 保水，锁水
water-lowering 降水量
compact concrete plate 紧凑型混凝土板
trenches with slope 沟坡
architectural outlining 建筑概述；外形轮廓
overground *adj*. 地上的
parallel slit wall 平行夹缝墙
monolith *n*. 独块巨石，整体塑制品
envisage *vt*. 想象，设想；正视，面对
hydraulic jack 液压起重器；液压千斤顶
Blvd.(=boulevard) *n*. 大马路；林荫大道
face chamber 面密室
terrain *n*. 地形，地势；地面
lining *n*. 衬里，里子；衬料；衬砌
formwork *n*. 建筑用模子材料
inherent strength 固有强度
steel arc frame 钢弧框架

shotcrete n. 喷混凝土
cross section 横截面；横断面；横剖面
shallow deployment 浅层部署
ventilation system 通风系统

【Translation】

Geotechnical engineering characteristics and general construction technology 岩土工程特点及一般施工技术

Geotechnical engineering construction has the characteristics of concealment, complexity and rigour. At present, the application of geotechnical engineering construction technology mainly includes: foundation treatment, foundation engineering construction technology, the slope reinforcement engineering construction technology, trenchless technology and geotechnical engineering testing technology, etc. Some of the typical construction technologies are discussed as follows.

岩土工程施工具有隐藏性、复杂性以及严格性的特点。目前而言，我国岩土工程施工技术应用比较广泛的主要有：地基处理技术、基础工程施工技术、边坡加固工程施工技术、非开挖技术以及岩土工程测试技术等。下面将介绍一些典型的施工技术。

Diaphragm walls 隔水墙/地下连续墙

Diaphragm walls bear with its static function the ground and water pressure and transfer them into the ground, through anchoring or ground resistance in its lower part. First of all, the excavation pit shall be made, in which guide walls are set up. They represent a kind of template for the diaphragm wall and define construction pit shape, as shown in Fig. 6.1. Furthermore, they ensure correct direction of grapple and stabilize the upper zone of diaphragm wall. The diaphragm wall is executed in panels. **In the course of** diaphragm wall construction a bentonite suspension is used, which strengthens the ground through its hydraulic pressure. When advancing with the excavation works and as a consequence of suspension loss in the underlying ground, **it is necessary constantly to be refilled**. After the required depth is reached, reinforcement package is lowered and then the panel is concreted with concreting pump.

地下连续墙利用静态功能承受着土压和水压，并通过锚固或较低位置的地基阻力将它们转移至地下。首先，要做基坑开挖，并在其中设置导墙。它们代表的是一种用于地下连续墙的样板，并以此确定施工基坑的形状，如图6.1所示。此外，导墙确保抓斗的正确方向并稳定地下连续墙的上部区域。地下连续墙分块施工。在地下连续墙施工过程中，会使用一种膨润土悬浮液，它通过液压加固地基。当开挖工作向前推进时，因为悬浮液会流失至下面的地层，有必要不断地进行补充。当达到所需的深度后，放下加固包，然后用混凝土泵将该块浇筑成墙。

This method treats the execution of diaphragm walls with special diaphragm wall equipment.

这种方法需使用专用地下连续墙设备完成地下连续墙的施工。

The diaphragm walls are formed out of reinforced concrete elements, executed one by one in the ground. This type of construction is able to perform support-, reinforcement-and water resistant functions, which determines their application field. The most frequent use of diaphragm walls is the excavation strengthening in urban conditions.

地下连续墙是由钢筋混凝土构件构筑而成，在地下一块一块地进行浇筑。这种类型的施工是能够发挥支护、加固和防水的功能，这些功能决定了其应用领域。地下连续墙最常见的用途是在城市环境中的开挖加固功能。

Anchoring of the ground 地锚

The fundamental concept of ground reinforcement with ground nails involves fortification of the ground through passive driving in of nails, close to each other, in order to create a coherent construction and thus increase the overall strength of ground slitting and to limit dislocation. The main idea is to transmit sustainable tensile strength, generated by tamping the ground through friction, created in the surfaces.

用地钉加固地基的基本理念包括：为了构建相连贯的建筑并进而增加地基分切后的整体强度以及限制错位，通过被动打入相互毗邻的地钉进行地基强化。主要思想是通过接触面间产生的摩擦力使地基得到夯实，进而转换成可以承受的抗拉强度。

Self-drilling IBO anchors are successfully used for anchoring in weak soil, coherent and non-coherent soils, and in case of unstable drilling.

自动钻进式的 IBO 锚杆被成功地用于软土、黏性土和非黏性土，以及无法稳定钻进条件下的锚固。

Self-drilling anchors are installed with drilling equipment with air or hydraulic rotation strikes, with the help of drilling wash, appropriate for specific soil conditions.

自动钻进式锚杆装配有空气或液压旋转钻头的钻井设备，在水冲钻探的帮助下，可以适用于特定的土壤条件。

Main applications:

主要应用:

-Tunnel digging

隧道掘进

- Reinforcement with rock bolts
  用岩石锚杆加固
- Anchoring of chambers
  硐室锚固
- Reinforcement in front of the tunnel chamber bolts
  隧道硐室工作面加固用锚杆
- Pilot with widening of the heel
  跟部加宽的导杆
- Tunnel portals, trenches and zones where the open method of construction is applied
  运用明挖式施工方法的隧道洞门、壕沟和其他地点

-Special activities

特殊活动

- Reinforcement of excavation pits, dams and swaths with injection anchors
  用喷锚进行加固的开挖基坑、大坝和其他地带
- Anchoring of support walls and noise barriers
  支撑墙和隔音板的锚固
- Pilot fundaments and control for their support in water—injection pilots/micropilots
  先导基础及其水中支护的控制——喷射先导/微型先导

Pilots/Drilling pilots   导洞/钻孔导洞

Building by means of pilot holes is one of the oldest methods. It is applied in soft or water-saturated soil or in narrow building sites. Pilots vary according to the type of building, disposition and method used.

用导洞进行建筑施工是最古老的方法之一。它适用于软土或水饱和的土或狭窄

的施工场地。导洞根据建筑施工的类型、配置和所使用的方法而变化。

In the process of building they are specified as single, tangent and cut-off pilots.

在建筑施工的过程中,它们被划分为单导洞、切线导洞和分段导洞。

Reinforced packages are prepared **in advance** and installed in one piece or **sequentially**, and then they **are covered with** concrete **in the drilled opening**. Cut-off pilots are placed in the following order: reinforced—not reinforced.

事先预备好加固包,安装在一起或按顺序安装,然后用混凝土将其浇固在钻开的地方。分段导洞按照以下顺序进行放置:加固层—不加固层。

Sheet piling 钢板桩

The sheet walls could be pressed, driven by vibrations or by hammering into the ground in compliance with the local ground characteristics. Moreover, they could be installed as dense wall, in the excavated ground diaphragm panels. Since sheet walls are executed as barriers, which are water impermeable, it is possible to meet the need of lowering ground water table on big areas through their joining with cohesive soil layer or artificial dense bottom.

根据当地地基的特点,钢板墙可以借助振动或打击的方法被压入或钻进地基中。此外,它们可以作为密实墙被安装在开挖好的地下连续墙板中。由于钢板墙可以做成不透水的屏障,通过与黏性土层或人工密实底板联合起来,可以满足大范围内降低地下水位的需要。

Drainage 排水系统

This system uses a series of wells, connected into a joint suction collector. Many single wells can **be connected to** one pump only (usually located in the suction collector). This type of system works effectively in variable soils and in shallow water-carrying layers, where the water level has been lowered to a waterproof layer, as shown in Fig. 6.2.

排水系统采用了一系列的井,并连接形成一个联合吸收器。许多单井只能连接到一个泵上(通常位于吸收器里面)。这类系统可在不同的土壤和浅的含水层中有效地发挥作用,其中的水位已降低到防水层的位置,如图 6.2 所示。

Construction excavations 施工开挖

Construction pits are formed from excavating the soil and include basements and the underground floors when the closed method of construction is applied. This includes all elements like walls, bottoms, reinforcement and retaining water. If the underground water level is above the lowest part of the pit (excavation bottom), they have to be lowered. This is done with the help of the so-called water-lowering by means of wells. Another method is to lay concrete on the excavation bottom in the shape of a compact concrete plate made of waterproof concrete. The excavation of foundation pit is shown in Fig. 6.3.

建筑基坑由开挖土壤而形成，包括采用封闭式施工方法时的地下室和地下楼层。这包括所有组成部分，如墙体、底部、加固与止水工程。如果地下水位高于基坑（开挖底部）的最低位置，地下水位必须设法降低。这可利用井的所谓"降低水位"功能的帮助来实现。另一种方法是将由防水混凝土制成的圆形混凝土板放在开挖基坑底部。基坑开挖如图 6.3 所示。

Methods for tunneling construction 隧道施工方法

The following methods are used in tunnel construction：

隧道施工中用到以下一些方法：

（1）Cut and cover method. Cut and cover method in reinforced trenches and in trenches with slopes respectively, depending on the presence of buildings nearby the route, or the presence of free space for digging trenches with slopes. The construction of the metro stations or the tunnels is implemented in these trenches. After that they are backfilled and the surface above is restored, as shown in Fig. 6.4. In the so built constructions installing of the equipment and architectural outlining is done. Massive reinforcements with great extent of hardness are mainly applied, thus allowing for the construction works in close proximity to the existing buildings and facilities without compromising their integrity. In some cases these reinforcements are parts of the supporting structure of the stations and the tunnels. The first station from the section St. Nedelja Sq. to housing complex Ljulin and three of the stations from the section St. Nedelja Sq. to Mladost, were built in a similar way. The same method is planned to be used for some of the stations from the second metro-diameter—Central Railway Station, Lavov most, National Palace of Culture and St. Naum.

（1）明挖法。明挖法可分别采用加固开挖槽和放坡式开挖槽的方式，取决于开挖路线附近存在的建筑物，或放坡式开挖可用的自由空间。地铁车站或隧道的建设是在这些开挖槽中实现的。之后，这些开挖槽被回填，而且地面得以重新恢复，如图 6.4 所示。在以这种方式建造的构筑物中，设备的安装及其外形轮廓也算完成了。主要使用具有极大硬度的大规模加固，因而能够允许在邻近已有建筑物和设施的地方进行不损坏其完整性的建筑施工。在一些情况下，这些加固成为车站和隧道支护结构的组成部分。从 St. Nedelja 广场至 Ljulin 住宅综合体区段的第一座车站以及从 St. Nedelja 广场到 Mladost 之间区段的 3 座车站，都是用相似的方法进行施工建造的。计划将同样的方法用到二号地铁径向部分的一些车站——中央火车站、Lavov most 站、国家文化宫站和圣瑙姆站。

(2) **Milan method.** Milan method is applied when a rapid restoration of the surface above the metro facilities is necessary. The Milan method allows for the simultaneous building of the underground and overground parts of the facilities. This approach is described as "cover and cut/dig", which means parallel slit walls, or pilots/sheet pile walls built and then connected to a monolith plate over them, after which the soil is excavated, as shown in Fig. 6.5. The Milan method comprises both building slit walls and pilots under the base and the fundamental system, prior to launching the tunnel excavation. Concrete slit walls, built with the help of bentonite solution, are used for enclosing walls. They have a double function: side support during the construction process and construction walls of the facility. This method has been applied partly in the construction of Joliot Curie and G. M. Dimitrov stations, as well as in some parts of the tunnels close to these stations. In the second metro-diameter, in order to minimize the time for stopping or to limit the traffic on the main streets, this method is envisaged to be applied at MS-8-II in front of TSUM, and MS-11-II under Cherny vrah Blvd., in Lozenets, as well as for the facilities to change the direction of movement after it.

(2) 米兰法。当地铁设施上方的地面需要得到快速恢复的情况下适合采用米兰法。米兰法允许对设施的地上和地下部分同步进行施工。这种方法被称为"盖挖法"，这意味着平行分割墙、或导洞/钢板桩墙构建后，在其上面连接成一整块板，此后再进行土方开挖，如图 6.5 所示。在进行隧道开挖前，米兰法包括构建分割墙、底板和基础体系下的导洞。利用膨润土溶液建成的混凝土分割墙被用来构建封闭式墙体。它们具有双重功能：施工过程中的侧向支护和设施本身的承重墙。在 Joliot Curie 和 G. M. Dimitrov 地铁站的建设工程以及邻近这些车站的隧道的部分工程中，该方法已经获得部分应用。在二号地铁径向段，为了使主要街道交通禁用或限用的通行时间最小化，考虑将该方法用到 TSUM 前方的 MS-8-II 工区，Lozenets 区 Cherny vrah 大道下的 MS-11-II 工区，以及之后为改变运行方向而建的设施工程。

(3) **Shield method.** The mechanized shield method is applied in the construction of tunnels in the central part of the city, where the tunnels are of considerable length; where archaeological sites are present, as well as to avoid open excavation works of considerable length along the major boulevards. The method involves mechanized excavation of soil in the front part of a steel cylinder, called "shield", and assembling the tunnel construction under the protection of the back

part of this cylinder. The machine is pushed forward by a system of hydraulic jacks, stuck in the tunnel construction in the back part of the shield. Because of the considerable depth of tunnels, this method was applied to the section St. Nedelja Sq. - V. Levski Stadium and it is intended to be used for the tunnels in the section Road junction Nadezhda - Han Asparuh Str. before reaching Patriarch Evtimij Blvd. Due to the location of the tunnels under the level of underground waters, shields with hydraulic counter pressure in the face chamber will be used for both sections. This will prevent sinking of the ground over the tunnels and the negative effect on the building above. The shield machine, which is 85m long and 1,600t heavy, will dig 9m daily and will transport about 1,000m^3 or 50 trucks of excavated material. The structure of shield machine and excavated shield tunneling are shown in Fig. 6. 6.

（3）盾构法。机械化的盾构法应用于城市中心区域的隧道建设中，尤其是当隧道足够长、存在古遗址，或者为了避免沿着主要大道进行长距离开挖。该方法包括位于被称为"盾构"的钢筒前部的机械化土体开挖，以及在该钢筒后部保护下进行隧道结构组装。整个机器由一个液压千斤顶系统推动前行，千斤顶位于盾构机后部的隧道组装施工段。由于隧道有足够的长度，该方法被用到 St. Nedelja 广场站至 V. Levski 体育馆站的区段，而且打算用在到达 Patriarch Evtimij 大道之前的 Nadezhda 与 Han Asparuh 街交叉路段的隧道施工中。由于隧道位置低于地下水位，工作面工作间带有液压平衡的盾构将用于这两个区段的施工建设。这将防止隧道上方岩土的下沉以及对上部建筑物的负面影响。盾构机长 85m，重 1600t，将每天挖进 9m，运出约 1000m^3 或 50 卡车的开挖料。盾构机结构及开挖的盾构隧道如图 6.6 所示。

(4) New Austrian tunnelling method. This method is applied effectively to sections with appropriate deployment of the terrain and limited water flow. It involves working on small sections of the face chamber stage by stage, then a primary construction/lining with reinforced cement solution is deployed above it and after that at certain distance a secondary construction/panelling of the tunnel is being reinforced and covered with concrete by means of a special movable formwork. The essence of the "New Austrian tunnelling method" implies transforming rock masses into support elements, as shown in Fig. 6. 7. It relies on the inherent strength of the surrounding rock mass being conserved as the main component of tunnel support. The tunnel is usually dug in a full profile; however it can be excavated in parts. The main idea is through embedding deformable supportive elements—anchors, steel arc frames and shotcrete—to form an alleviating arc to turn it into a support element, substituting the lining. The new stable equilibrium is determined by fading of the radial deformations of the excavation, measured on the spot. This method is applied in masses with random strength with sufficient tunnel covering to form the alleviating arc, and with tunnels with random size of the cross section.

This method was applied to the link between V. Levski Stadium and the tunnel under Dragan Tsankov Blvd. in front of the National Radio. Due to the specifics of the geologic conditions, deep ground waters and significant depth of deployment of the tunnels from the second metro-diameter in the section under Cherny vrah Blvd. between Hemus Hotel and James Bourchier Blvd., the new Austrian method will be applied to their construction. In the section after the metro station under Cherny vrah Blvd., due to the significant width of the facilities, shallow deployment, high level of ground water, and the presence of ventilation system facilities above it, the application of this method is not appropriate.

（4）新奥法。此方法有效地应用于具有适当土体条件和受限水流的区段。它涉及工作面室的小区块的分步施工，然后在上方用钢筋水泥溶液进行初次浇筑或衬砌，相隔一定距离后，利用专用可移动式模板，完成隧道的加固与混凝土浇筑施工或镶板。"新奥法"的实质是指将岩体本身转化为支护结构的组成部分，如图6.7所示。它依赖于作为隧道支护主体的围岩体所具有的固有强度。隧道通常是全断面开挖的，但也可以分步（块）开挖。主要思想是通过嵌入可变形的支护构件，比如锚杆、弧形钢架和喷射混凝土，由此形成一个缓冲拱，并转变成支护体，代替衬砌。新的稳定平衡状态可通过现场测量的因开挖引起的径向变形的逐渐消失来决定。该方法适用于具有足以形成缓冲拱的隧道覆盖层的任意强度的岩体条件，也适用于任意大小横截面的隧道。这种方法曾应用于V. Levsk体育馆和国家广播电台前方Dragan Tsankov大道下方隧道之间的连接线施工。位于Hemus酒店以及James Bourchier大道之间Cherny vrah大道下面地段的二号地铁径向隧道，考虑到其特殊的地质条件、深部地下水以及隧道布线的大埋深，将使用新奥法进行建设。在Cherny vrah大道下的地铁站之后的区段，由于设施的大宽度、浅的布线、高的地下水位以及在其上面存在通风系统设施，则不适合采用这种方法。

The selection principles of geotechnical engineering construction technology 岩土工程施工技术的选用原则

（1）**Economic principle**：Because of the "uncertainty" of construction technology, there are usually several sets of technology available for each type of geotechnical engineering problems. Then it needs to make comparison between the factors such as economy, time limit for a project, technology, safety etc. But in any case, the economy of technology will always dominate the selection.

（1）经济性原则：由于施工技术的"不确定性"，每类岩土工程问题往往有几套技术方案可供选择，因此需要进行经济、工期、技术、安全等方面的对比，但无论如何，技术的经济性将总是主导选择。

（2）**Applicability principle**：There is no absolute good or bad geotechnical engineering technology, i.e. a geotechnical technology should not be judged to be absolutely good or bad only by some/several indicators of the technology. It's not necessary to use the best, but we must use the most appropriate construction technology.

（2）适用性原则：在岩土工程中没有绝对好或不好的技术，亦即不能以某种技术的某个/几种指标来判定它是绝对好或差的技术。不一定使用最好的，但一定要使用最合适的施工技术。

（3）**Practical principle**：Because of the "uncertainty" of geotechnical engineering construction technology, the technical feasibility of construction technology must not be judged only by theoretical analysis and calculation. It's more important that the real engineering practice should be used to choose the most suitable method.

（3）实践性原则：由于岩土工程施工技术的"不确定性"，所以施工技术的可靠性不能只依靠理论分析和计算来判断，更重要的是运用具体的工程实践来选择最适合的方法。

（4）**Environmental protection principle**：With the country's increasingly stringent environmental requirements, the environmental effect in the geotechnical engineering construction and the influencement degree of various construction methods to the environment will become an important reference to construction technology selection. More and more environmental protection construction technologies will be the first choice of geotechnical engineering construction.

（4）绿色性原则：随着国家日益严格的环保要求，岩土工程施工中的环境效应以及各种施工方法对环境的影响程度将成为施工技术选用的重要参考。越来越多的绿色环保施工技术将成为岩土工程施工的首选。

【Important sentences】

1. In the course of diaphragm wall construction a bentonite suspension is used, which strengthens the ground through its hydraulic pressure.
   In the course of 在……的过程中。
2. it is necessary constantly to be refilled.
   有必要不断地进行补充。constantly 表"持续地、不断地"。
3. Reinforced packages are prepared in advance and installed in one piece or sequentially, and then they are covered with concrete in the drilled opening.
   in advance 预备、预先；sequentially 副词，表"继续地、循序地"；be covered with 覆盖；in the drilled opening 作位置状语，在钻洞的地方。
4. Many single wells can be connected to one pump only.
   be connected to 连接到。

6.2 Construction equipments 施工设备

【Text】

Excavation and loading

One family of construction machines used for excavation is broadly classified as a crane-shovel, which consists of three major components, shown as in Fig. 6.8.

(1) A carrier or mounting which provides mobility and stability for the machine.

(2) A revolving deck or turntable which contains the power and control units.

(3) A front end attachment which serves the special functions in an operation.

Fig. 6.8 Three major components of construction machine
图 6.8 施工机械的三大组成构件
1—carrier; 2—revolving deck; 3—front end attachment
1—装载设备; 2—旋转式甲板; 3—前端附件

Crane

A crane is a type of machine, generally equipped with a hoist, wire ropes or chains, and sheaves, that can be used both to lift and lower materials and to move them horizontally, as shown in Fig. 6.9 and Fig. 6.10. It is mainly used for lifting heavy things and transporting them to other places. It uses one or more simple machines to create mechanical advantage and thus move loads beyond the normal capability of a human. Cranes are commonly employed in the transport industry for the loading and unloading of freight, in the construction industry for the movement of materials and in the manufacturing industry for the assembling of heavy equipment.

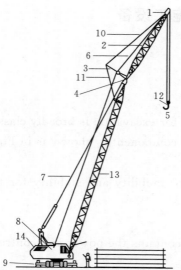

Fig. 6.9 A modern mobile crane with outriggers
图 6.9 带有起重臂的现代移动式起重机
1—upper sheave; 2—jib; 3—jib strut; 4—boom point; 5—hook; 6—hoist rope; 7—boom guy line; 8—winding drum; 9—outrigger; 10—jib guy line; 11—jib backstay; 12—lower sheave; 13—latticed boom; 14—slewing platform
1—上滑轮；2—臂架；3—吊臂支撑；4—伸臂末端；5—吊钩；6—起升钢丝绳；7—吊杆绷绳；8—提升绞筒；9—起重臂；10—臂架绷绳；11—动臂桅杆；12—下滑轮；13—笼格吊杆；14—回转台

Control costs

6 Ways equipment data can help cut equipment costs (from CAT).

Using the data generated by your equipment puts money in your pocket by helping you keep your equipment in top shape and avoid unplanned downtime. Here are six ways to make the most of your data to help manage your equipment and reduce operating costs.

(1) **Preventive maintenance.** Take advantage of electronic alerts to schedule and complete all recommended maintenance and service.

(2) **S·O·SSM fluid analysis.** Sample oil, coolant and hydraulic fluids regularly, have your CAT dealer analyze them for wear indicators, then act quickly on the results and recommendations from your dealer's Condition Monitoring Analyst (CMA).

(3) **Inspections.** Regularly look for smoke, leaks, lagging performance and so on. Send electronic inspection reports to your CAT dealer for use in a condition monitoring program.

(4) **Repair before failure.** Use equipment data and electronic alerts to catch small problems early. Schedule repairs quickly to avoid breakdowns and keep repair costs low.

(5) **Operator training.** Operating practices dramatically impact performance and component wear. Remotely monitor your operators' day-to-day performance to spot training opportunities and reward good performance.

(6) **Record keeping.** Automatically gather data on machine history, component life and operating costs. Good records help you identify high-cost or problem areas, track work flow, control expenses and increase machine resale value.

6.2 Construction equipments 施工设备 /153/

(a) Crawler crane
履带式起重机

(b) Tower crane
塔吊

(c) Truck mounted crane
车载式起重机

Fig. 6.10 Crane and tower crane
图 6.10 吊车和塔吊

CAT equipment

CAT equipment sets the standard for our industry. The CAT product line of more than 300 machines reflects our increased focus on customer success. We will remain the leader by continuing to help our customers meet their needs with durable and reliable equipment. Caterpillar has the best distribution and product support system in any capital goods industry.

【Key words】

crane-shovel 起重机铲
carrier n. 搬运人；运输公司；搬运器
mounting n. 装备；衬托纸
revolving deck 旋转式甲板
turntable n. 转盘；转台
front end attachment 前端附件

hoist *n*. 起重机，升降机；升起
wire rope　钢缆，钢索
sheaves（sheaf 的复数形式）*n*. 捆，束，扎
freight *n*. 货运，货物；运费
outrigger *n*. 舷外支架，突出的梁，桁
latticed *adj*. 装有格子的
downtime *n*. （工厂等由于检修、待料等的）停工期
coolant *n*. 冷冻剂，冷却液，散热剂
hydraulic fluid　液压机液体
CAT　Caterpillar 公司的简称
dealer *n*. 商人；发牌人；经销商
breakdown *n*. 崩溃，倒塌；损坏，故障
component wear　部件磨损

【Translation】

Excavation and loading　挖掘与装载

One family of construction machines used for excavation is broadly classified as a crane-shovel, which consists of three major components, shown as in Fig. 6.8.

有一类开挖施工机械被大致归类为起重机铲，由三大构件组成，如图 6.8 所示。

(1) A carrier or mounting which provides mobility and stability for the machine.

为机器提供机动性和稳定性的装备。

(2) A revolving deck or turntable which contains the power and control units.

包含动力和控制单元的旋转式甲板或可转动平台。

(3) A front end attachment which serves the special functions in an operation.

在使用中起特殊作用的前端附件。

Crane　吊车/起重机

A crane is a type of machine, generally equipped with a hoist, wire ropes or chains, and sheaves, that can be used both to lift and lower materials and to move them horizontally, as shown in Fig. 6.9 and Fig. 6.10.. It is mainly used for lifting heavy things and transporting them to other places. It uses one or more simple machines to create mechanical advantage and thus move loads beyond the normal capability of a human. Cranes are commonly employed in the transport industry for the loading and unloading of freight, in the construction industry for the movement of materials and in the manufacturing industry for the assembling of heavy equipment.

起重机是一种机器类型，一般都配备有绞车、钢丝绳或链条和滑轮，它可用于提升或放下材料并使它们完成水平移动，如图 6.9 和图 6.10 所示。它主要用于提升重物并把它们搬运到其他地方。它使用一个或多个简单的机器来创造出机械的优势，从而超越人类的正常能力来移动重物。起重机通常在运输行业中用于货物的装卸，在建筑行业中用于建材的运送，在制造业中则用于重型设备的组装。

Control costs　控制成本

6 Ways equipment data can help cut equipment costs (from CAT).

6 方面的设备数据可以帮助降低设备成本（来自 CAT）。

Using the data generated by your equipment puts money in your pocket by helping you keep your equipment in top shape and avoid unplanned downtime. Here are six ways to make the most of your data to help manage your equipment and reduce operating costs.

利用设备生成的数据可以使设备保持最佳状态，避免意外停机，实现节约用钱。这里介绍六种可以最大限度地利用设备数据、帮助管理设备和降低运营成本的方法。

(1) **Preventive maintenance. Take advantage of** electronic alerts to schedule and complete all **recommended** maintenance and service.

(1) 定期检修。利用电子警报来安排和完成所有推荐的维护和服务。

(2) **S·O·SSM fluid analysis.** Sample oil, coolant and hydraulic fluids regularly, have your CAT dealer analyze them for wear indicators, then act quickly on the results and recommendations from your dealer's Condition Monitoring Analyst (CMA).

(2) S·O·SSM 流体分析。定期对润滑油、冷却液和液压油取样，请 CAT 经销商分析其磨损指标，然后对经销商的状态监测分析（CMA）结果和建议进行快速处置。

(3) **Inspections.** Regularly look for smoke, leaks, lagging performance and so

on. Send electronic inspection reports to your CAT dealer for use in a condition monitoring program.

（3）检查。时常关注烟雾、泄漏、滞后性能等问题。将电子检查报告发给 CAT 经销商，用于状态监测计划。

（4）**Repair before failure.** Use equipment data and electronic alerts to catch small problems early. Schedule repairs quickly to avoid breakdowns and keep repair costs low.

（4）故障前维修。使用设备数据和电子警报及早发现小问题。为避免死机并使维修成本低，需迅速安排维修。

（5）**Operator training. Operating practices dramatically impact** performance and component wear. Remotely monitor your operators' day-to-day performance to spot training opportunities and reward good performance.

（5）操作员培训。实际操作显著影响机器的性能和部件的磨损。远程监控操作员的日常表现，从中发现培训机会，而且对好的表现予以奖励。

（6）**Record keeping.** Automatically gather data on machine history, component life and operating costs. Good records help you identify high-cost or problem areas, track work flow, control expenses and increase machine resale value.

（6）保存记录。自动收集机器使用历程、部件寿命和运行费用的数据。良好的记录能帮助你识别高成本或出现问题的地方，也能追踪工作流程、控制开销和提升机器转售价值。

CAT equipment　　CAT 设备

CAT equipment sets the standard for our industry. The CAT product line of more than 300 machines reflects our increased focus on customer success. We will remain the leader by continuing to help our customers meet their needs with durable and reliable equipment. Caterpillar has the best distribution and product support system in any capital goods industry.

CAT 设备设定了行业标准。超过 300 种机器的 CAT 生产线反映出越来越关注客户的成功。我们将继续通过用耐用而可靠的设备帮助客户达成需要的方式，保持行业领先地位。卡特彼勒公司在任何重型设备行业都拥有最佳的经销和产品支持体系。

【Important sentences】

1. Take advantage of electronic alerts to schedule and complete all recommended

maintenance and service.

Take advantage of 固定短语，利用；recommended 过去分词作定语，推荐的。

2. Operating practices dramatically impact performance and component wear.

Operating practices 动名词作主语；dramatically impact 显著地影响。

6.3 Construction cost and safety 工程建设成本与安全性

【Text】

Cost savings through better safety

Past historical data gives proof that investing in safety and health will help control a company's bottom-line during all phases of the company life cycle. The one problem with this statement is that society today is all about the "here and now" and is often not concerned with the future. In many situations, employers worry about safety and health only as it relates to being regulatory compliant at a minimum with governing agencies. But if these companies fully embrace the concepts of implementation and make safety and health a core value, the potential return on investment relating to injury prevention, along with the creation of a safe work environment can be huge, monetarily and culturally.

Return of investment from making safety and health a core value includes the following:

- Better safety efficiency (pre-planning) without creating higher costs and schedule interruptions
- Employee injuries accounting for only 2.5 percent of project costs because of a well-implemented safety program compared with 6 to 9 percent of project costs without such a program
- Lower injury rates, which in turn means higher profit margins. "Safety is a function of management" (Heinrich, 1959; Murphy, 1992; Brauer, 1994)

Throughout history, data shows this statement to be true. Improved safety practices have reduced injuries and fatalities by more than 50 percent in the past three decades. This 50 percent drop in worker injuries contradicts the fatalistic view by management in the mid-1960s that levels of injuries were relative to levels of hazards in the work environment, and that the injuries were unavoidable.

Safety should always be viewed in a positive view-not in a fatalistic manner. A company's main objective should be to have no worker injuries. With this objective must be an unbreakable belief that the goal of zero injuries is attainable. While companies should plan to have no injuries, they must still have a well thought-out safety plan in the event of a work-related injury. This is where the problem lies. How can safety professionals justify spending money on a theory or a hunch that an accident or incident might happen to the company and its employees. Most companies look at expenses versus revenue and just focus on tangible elements of business and not on possibilities.

Employees do not have total control of their safety, because a safe work environment consists of more than just employee involvement. Some factors to consider changing include the use of funds to make physical conditions safer and improved coordination of work activities (Hinze, 1997). These two factors rely heavily on managerial personnel to identify unsafe conditions and less on unsafe employee work habits.

In general, accident prevention is accepted as the responsibility of management (Hinze, 1997). However, companies differ in the amount of resources they dedicate to accident prevention programs. This difference could stem from varying perceptions of the degree of influence that management can have on reducing worker injuries. "If the true costs of injuries were well defined, management would be in a better position to make informed decisions concerning safety. Rather than addressing safety solely from an altruistic point of view, owners should also consider safety from a more purely economic perspective." (Hinze, 1997)

Obviously, safety should be a priority and responsibility of management simply because managers are concerned about the welfare of those who work for them, but in addition, there are the out-of-pocket costs incurred when injuries occur. Research data collected by the Department of Labor has shown that between 1980 and 1987, workers' compensation insurance cost doubled in the United States. This upward trend has continued. The Workplace Safety and Insurance Board (WSIB) forecasted that insurance premiums will increase from 5 to 30 percent in the next year, depending on the state, due to the amounts being claimed. Companies could be paying 10 to 20 percent of their direct labor cost for workers' compensation premiums due to higher Experience Modification Rates (EMR), which in turn exceeds their profit margins.

The Department of Labor further indicates that it is not uncommon for companies with poor safety performance to pay twice the premium costs compared to those who have good safety records. Increasing medical costs, litigation and ballooning insurance premiums that are identifiable costs associated with an accident are considered direct costs. One problem when identifying the direct costs associated with safety accidents is that indirect costs or hidden costs are also going to be conjoined with direct costs. The magnitude of the accident, not necessarily the severity of the injury, makes the ratios of direct and indirect costs vary greatly. Some estimates of ratios between indirect and direct costs vary from about 1∶1 to 20∶1 (Department of Labor).

To better understand direct and indirect cost, direct costs are those costs directly attributed to or associated with an injury (Hinze, 1997). Most of the direct costs are usually covered by the workers' compensation insurance policies. Some examples of direct costs include:

- Any type of transportation/ambulance service required for the injured party
- Medical and ancillary treatment prescription by a healthcare provider
- Prescribed medication costs
- Hospitalization
- Disability benefits, which can include lost wages of the injured person

Thus, the direct cost of injuries tend to be those associated with the treatment of the injury and any unique compensation offered to workers as a consequence of being injured (Hinze, 1997).

In 2001, the Workplace Safety and Insurance Board (WSIB) identified the cost of one lost-time injury to cost an employer an average of $35,000 in direct costs alone.

Indirect costs are more elusive than costs associated with an employee accident, and are often referred to as hidden costs. Indirect costs are costs with no retrieval mechanism to accurately associate the cost to the injury-the idea being that it is not a matter of whether a specific cost has been identified, but whether the cost is a result of the injury.

In the construction industry, indirect costs of injuries are much easier to detect due to the different work environment construction workers are involved in on a daily basis. When a construction worker is injured, treatment is immediately given, either first aid or medical attention, depending on the severity. During this time, the injured worker is being paid for his or her time, travel time and work crew stoppage. And when he returns to work, he will not be working at 100 percent. Also, if the injury is particularly severe or dramatic, supervisors and company officials may need to be involved because of media attention over the accident.

With the growing costs of the industry, safety professionals have acknowledged the presence of indirect, or hidden, costs of injuries.

Many researchers, when studying the concepts of direct and indirect cost, try to express their relationship in a ratio of indirect costs to direct costs. H. W. Heinrich in 1979 was the first to try this, and he estimated the ratio of the indirect costs of injuries to the direct costs to be approximately 4∶1. The data used to reach this conclusion was information gathered from various industrial facilities in the United States. The one problem with Heinrich's 4∶1 ratio was that it did not have a chance to factor in the steep escalation of the health care costs (direct costs), which might reasonably be expected to be lower today.

Heinrich's factor of four was stated for many years as the ratio of the indirect costs of injuries to the direct costs. Computations in the 1980s became less conservative. In a study conducted by R. Sheriff (1980), he showed that the ratio of indirect and direct costs may be as high as 10 compared to Heinrich's four. F. Bird and R. Loftus (1976), tracked the ratio to be as high as 50. With all these studies representing different ratios, the common bond accepted was that indirect costs were very significant when examining the cost of an injury.

Results of historical studies indicate that greater attention should be given to indirect costs of worker injuries. The data from indirect costs of medical-case injuries are nearly similar to the direct costs. Even with the data from indirect and direct costs being very similar in amounts, indirect costs balloon exponentially when liability cases are sought for a worker's injury. Very large monetary amounts can still accumulate into large sums when there are multiple injuries. With this understanding, here is a brief example of how much an injury can affect a company's output or bottom-line if trying to make a three percent profit goal:

If a company acquired a $50,000 loss due to injury, illness or damage and still tries to make a 3 percent profit, the company theoretically must increase sales of services by an additional $1,667,000. [Example taken from Electronic Library of Construction Occupational Safety and Health (ELCOSH)]

Even with this data there is still unwillingness to embrace safety and health programs fully within some organizations. This reluctance is mainly due to fear and the ability to maintain adequate safety personnel.

Studies show that there is direct correlation between lower accident rate reduction and increased safety investment, and increases safety investment in relation to increased profits. As noted earlier, the implementation of a safety and health program will not have a huge financial impact at the beginning stages of implementation, but given time, it will certainly pay for itself and make the company profit in years to come.

Relationship between construction safety and quality performance

It is well established that the project cost, quality, safety, and duration are the four critical elements that contribute to project success. Past literature has established theoretical relationships between construction safety and quality on the basis of opinions of industry experts. This is the first empirical inquiry into the relationship between safety and quality, testing the null hypothesis that there is no statistical relationship among quality performance indicators and safety performance indicators. To test this hypothesis, empirical data were collected from 32 building construction projects, ranging in scope from \$50,000 to \$300 million dollars. Several quality metrics (e.g., cost of rework per \$1M project scope and rate of rework per 200,000 worker-hours) were used as predictor variables and first aid and Occupational Safety and Health Administration (OSHA) recordable injury rates were used as response variables. Linear regressions among the predictor and response variables showed that there are two statistically significant relationships: the OSHA recordable injury rate is positively correlated to rework ($r^2=0.968$; p-value$=0.032$) and the first aid rate is positively correlated to number of defects ($r^2=0.548$; p-value$=0.009$).

To understand why these relationships exist and to identify specific strategies that promote both safety and quality, open-ended interviews were conducted with project managers. These individuals indicated that the most compelling reason for the strong positive correlation between rework and injuries is the fact that rework involves demolition, schedule pressure, and unstable work processes. They also noted that devoting resources to preplanning, allowing the necessary time to complete tasks correctly the first time, encouraging leadership at the workface, and encouraging workers to take pride in their work are all strategies that promote both safety and quality.

【Key words】

stem $vt.$ 遏制，阻止（液体的流动等）
altruistic $adj.$ 利他的，无私心的
elusive $adj.$ 难以捉摸的；不易记住的
retrieval mechanism 检索机制
escalation $n.$ 扩大，增加
exponentially $adv.$ 以指数方式

monetary *adj*. 货币的，金钱的；金融的，财政的
designate *vi*. 把……定名为，称呼委任，指派
S. M. ASCE 美国土木工程师协会学生会员
null hypothesis 虚假设，解消假设；无效（无价值）假说；零假说
preplanning *n*. 预先计划，前规划
CSP＝Certified Software Professional 认证软件工程师
CHST＝Construction Health & Safety Technician 建设健康及安全人员

【Translation】

Cost savings through better safety 抓好安全节约成本

Past historical data gives proof that investing in safety and health will help control a company's bottom-line during all phases of the company life cycle. The one problem with this statement is that society today is all about the "here and now" and is often not concerned with the future. **In many situations, employers worry about safety and health only as it relates to being regulatory compliant at a minimum with governing agencies.** But if these companies **fully embrace** the concepts of implementation and make safety and health a core value, the potential return on investment relating to injury prevention, **along with** the creation of a safe work environment can be huge, monetarily and culturally.

过去的历史数据证明，安全和健康方面的投资将帮助控制公司生命周期中所有阶段的公司底线（基本运营成本）。与此相关的问题在于，当今社会只注重"当下"而常常忽视未来。在许多情况下，雇主担心安全和健康只是因为它关系到要满足政府机构要求的最低的法制承诺。但如果这些公司完全接纳做到实处的观念，并把安全与健康作为一种核心价值，与工伤预防相关的投资，会随着创造出一个安全的工作环境，在金钱上和企业文化上带来潜在的巨大回报。

Return of investment from making safety and health a core value includes the following:

通过把安全和健康作为一种核心价值可获得如下几点投资回报：

- Better safety efficiency (pre-planning) without creating higher costs and schedule interruptions
  更好的安全效率（预先计划）而不产生更高的成本且不致中断进度
- Employee injuries accounting for only 2.5 percent of project costs because of a well-implemented safety program compared with 6 to 9 percent of project costs without such a program
  一个实施良好的安全计划，会使工伤仅占项目成本的2.5%，与此相比，

没有这样的计划，则会占项目成本的 6%～9%
- Lower injury rates, which in turn means higher profit margins. "Safety is a function of management" (Heinrich, 1959; Murphy, 1992; Brauer, 1994) 受伤率更低，反过来意味着更高的利润空间。"安全是一种管理职能"（海因里希，1959；墨菲，1992；布劳尔，1994）

Throughout history, data shows this statement to be true. Improved safety practices have reduced injuries and fatalities by more than 50 percent in the past three decades. This 50 percent drop in worker injuries contradicts the fatalistic view by management in the mid-1960s that levels of injuries were relative to levels of hazards in the work environment, and that the injuries were unavoidable.

纵观历史，数据表明这个说法是正确的。在过去的 30 年中，改良好的安全实践已经使伤亡减少了 50%以上。工伤下降 50%驳斥了 20 世纪 60 年代中期管理界宿命论的观点，即工伤水平跟工作环境的危害程度有关且是不可避免的。

Safety should always be viewed in a positive view-not in a fatalistic manner. A company's main objective should be to have no worker injuries. With this objective must be an unbreakable belief that the goal of zero injuries is attainable. While companies should plan to have no injuries, they must still have a well thought-out safety plan in the event of a work-related injury. This is where the problem lies. How can safety professionals justify spending money on a theory or a hunch that an accident or incident might happen to the company and its employees. Most companies look at expenses versus revenue and just focus on tangible elements of business and not on possibilities.

应该一直用一种积极向上眼光去看待安全问题，而不是用一种听天由命的方式去看待。一个公司的主要目标应该是没有工人受伤。有了这样一种牢不可破的信念，零伤亡的目标就是可以实现的。然而在公司制订计划保证没有人员受伤的同时，还须对因公受伤的事件有慎重且周到的安全计划。问题在于，安全事务专业人员怎样才能证明应该为事故或意外可能发生在公司及其员工身上的理论或预感花钱。因为大多数公司会关注开支与收入，而且只注重于那些摸得到的经营要素而不是可能性。

Employees do not have total control of their safety, because a safe work environment consists of more than just employee involvement. Some factors to consider changing include the use of funds to make physical conditions safer and improved coordination of work activities. These two factors rely heavily on managerial personnel to identify unsafe conditions and less on unsafe employee work habits.

员工不能完全掌控其自身安全,因为安全的工作环境包含的不仅仅是员工的参与。要考虑进行改进的一些因素包括使用专项资金使得硬件条件更安全以及改进工作活动间的协调性。这两个因素严重依赖于负责鉴定不安全条件的管理人员,而与不安全的员工工作习惯关系较轻。

In general, accident prevention is accepted as the responsibility of management. However, companies differ in the amount of resources they dedicate to accident prevention programs. This difference could stem from varying perceptions of the degree of influence that management can have on reducing worker injuries. "If the true costs of injuries were well defined, management would be in a better position to make informed decisions concerning safety. Rather than addressing safety solely from an altruistic point of view, owners should also consider safety from a more purely economic perspective."(Hinze,1997)

在一般情况下,事故预防被视为是管理的责任。然而,企业为致力于事故预防的计划分配资源却不同。这种不同源于对管理能减少工伤影响程度的看法上的差异。"如果工伤的真实成本能进行明确定义,管理层将可处在一个更好的位置来做出关于安全的明智决定。而不仅仅是根据利他行为的观点对待安全,公司所有者也可以从更纯粹的经济视角来看待安全。"(欣策,1997)

Obviously, safety should be a priority and responsibility of management simply because managers are concerned about the welfare of those who work for them, but in addition, there are the out-of-pocket costs incurred when injuries occur. Research data collected by the Department of Labor has shown that between 1980 and 1987, workers' compensation insurance cost doubled in the United States. This upward trend has continued. The Workplace Safety and Insurance Board (WSIB) forecasted that insurance premiums will increase from 5 to 30 percent in the next year, depending on the state, due to the amounts being claimed. Companies could be paying 10 to 20 percent of their direct labor cost for workers' compensation premiums due to higher Experience Modification Rates (EMR), which in turn exceeds their profit margins.

显然,安全应该优先,同时也是管理者的责任,因为管理者应关注那些为他们工作的人的福利,但除此以外,当有受伤发生时也会招致自掏腰包。由劳动部门收集的研究数据显示,1980—1987年间,美国工人的职工赔偿保险费用翻了一番。这种上升趋势一直在持续。工作场所安全与保险委员会(WSIB)预测,在接下来的一年中,基本保险费将增加5%~30%,取决于所在的州,由于各州索赔额不同。由于更高的经验修正率(EMR),公司可能要将直接人工成本的10%~20%用于支付工人的基本保险费,这反过来又超过了公司的利润空间。

The Department of Labor further indicates that it is not uncommon for companies with poor safety performance to pay twice the premium costs compared to those who have good safety records. Increasing medical costs, litigation and ballooning insurance premiums that are identifiable costs associated with an accident are considered direct costs. One problem when identifying the direct costs associated with safety accidents is that indirect costs or hidden costs are also going to be conjoined with direct costs. The magnitude of the accident, not necessarily the severity of the injury, makes the ratios of direct and indirect costs vary greatly. Some estimates of ratios between indirect and direct costs vary from about 1∶1 to 20∶1.

劳工部进一步表明，与那些有良好安全记录的公司相比，安全性能差的公司要支付双倍的保费成本并不罕见。日益增长的医疗费用、诉讼费用和不断膨胀的基本保险费等是与事故相关的能确认的成本，能作为直接成本对待。在确认与安全事故有关的直接成本时，存在间接成本或隐藏成本也会附带在直接成本里面的问题。事故的严重程度不一定是受伤的严重程度，会使直接成本和间接成本之间的比例变化很大。一些间接成本和直接成本之间的比率估值约从1∶1到20∶1之间变化。

To better understand direct and indirect cost, direct costs are those costs directly attributed to or associated with an injury. Most of the direct costs are usually covered by the workers' compensation insurance policies. Some examples of direct costs include：

为了更好地理解直接和间接成本，直接成本指那些直接归因于受伤的成本。大部分的直接成本通常由工人的职工赔偿保险政策负担。一些直接成本的例子包括：

- Any type of transportation/ambulance service required for the injured party
  为受害方提供所需的任何类型交通工具/救护车服务
- Medical and ancillary treatment prescription by a healthcare provider
  由保健机构提供的医疗和辅助治疗处方
- Prescribed medication costs
  处方药的费用
- Hospitalization
  住院治疗
- Disability benefits, which can include lost wages of the injured person
  伤残补贴，包括受伤人损失的工资

Thus, the direct cost of injuries tend to be those associated with the treatment of the injury and any unique compensation offered to workers as a consequence of being injured.

因此，受伤的直接成本大体是受伤的治疗费用加上作为赔付给因公受伤工人的所有专门补偿费用。

In 2001, the Workplace Safety and Insurance Board (WSIB) identified the cost of one lost-time injury to cost an employer an average of \$35,000 in direct costs alone.

2001年，工作场所安全与保险委员会（WSIB）发现一个失时工伤的成本仅在直接成本中就花费雇主平均35000美元。

Indirect costs are more elusive than costs associated with an employee accident, and are often referred to as hidden costs. Indirect costs are costs with no retrieval mechanism to accurately associate the cost to the injury-the idea being that it is not a matter of whether a specific cost has been identified, but whether the cost is a result of the injury.

间接成本比与员工事故所产生的成本相比更难以捉摸，它们通常被作为隐藏成本。间接成本是那些没有可遵循的机制来准确将成本与受伤关联起来的成本，因此，已经不是一个某项特定成本是否已被确认的问题，而是这一成本是否是受伤的结果的问题了。

In the construction industry, indirect costs of injuries are much easier to detect due to the different work environment construction workers are involved in on a daily basis. When a construction worker is injured, treatment is immediately given, either first aid or medical attention, depending on the severity. During this time, the injured worker is being paid for his or her time, travel time and work crew stoppage. And when he returns to work, he will not be working at 100 percent. Also, if the injury is particularly severe or dramatic, supervisors and company officials may need to be involved because of media attention over the accident. With the growing costs of the industry, safety professionals have acknowledged the presence of indirect, or hidden, costs of injuries.

在建筑行业中，由于所涉及的工作环境不同，施工人员以天为基础安排工作，所以受伤的间接成本更容易被测定。当施工人员受伤时，会立即给予处置，要么采取急救，要么求医，取决于严重程度。在此期间，将继续支付受伤工人工时费、差旅费以及工作团队误工费。而且当他回到工作岗位时，将不能百分之百地继续工作。此外，如果受伤情况特别严重，因为媒体对事故的关注也可能会使监督方和公司管理人员牵连其中。随着工业成本的不断增长，安全事务专业人士已经承认存在间接或隐藏的受伤成本了。

Many researchers, when studying the concepts of direct and indirect cost, try to express their relationship in a ratio of indirect costs to direct costs. H. W. Heinrich in 1979 was the first to try this, and he estimated the ratio of the indirect costs of injuries to the direct costs to be approximately 4∶1. The data used to reach this conclusion was information gathered from various industrial facilities in the United States. The one problem with Heinrich's 4∶1 ratio was that it did not have a chance to factor in the steep escalation of the health care costs (direct costs), which might reasonably be expected to be lower today.

许多研究者在研究直接和间接成本的概念时，尝试用间接成本和直接成本之间的比率来表示两者之间的关系。H. W. 海因里希在1979年首次进行了这一尝试，他估算受伤的间接成本与直接成本之比大约为4∶1。用于得出这一结论的数据来自从美国的各种工业企业收集到的信息。海因里希给出的4∶1这一比例存在一个问题，就是它没有机会考虑到医保费陡峭上升的趋势（直接成本），因此这一比率在今天可能是有所下降的。

Heinrich's factor of four was stated for many years as the ratio of the indirect costs of injuries to the direct costs. Computations in the 1980s became less conservative. In a study conducted by R. Sheriff (1980), he showed that the ratio of indirect and direct costs may be as high as 10 compared to Heinrich's four. F. Bird and R. Loftus (1976), tracked the ratio to be as high as 50. With all these studies representing different ratios, the common bond accepted was that indirect costs were very significant when examining the cost of an injury.

海因里希的四分说多年来一直用于表述直接成本与间接成本之间的比例。20世纪80年代的计算方法变得不那么保守了。在R. Sheriff（1980年）完成的一项研究中，他指出与海因里希的四分说相比，间接成本和直接成本之比可能高达10。F. Bird和R. Loftus（1976年）跟踪发现该比率可高达50。虽然上述研究给出了不同的比率，但可接受的共同点是在查验受伤成本时，间接成本是非常重要的。

Results of historical studies indicate that greater attention should be given to indirect costs of worker injuries. The data from indirect costs of medical-case injuries are nearly similar to the direct costs. Even with the data from indirect and direct costs being very similar in amounts, indirect costs balloon exponentially when liability cases are sought for a worker's injury. Very large monetary amounts can still accumulate into large sums when there are multiple injuries. With this understanding, here is a brief example of how much an injury can affect a company's output or bottom-line if trying to make a three percent profit goal:

历史研究的结果表明，应该给予工人受伤所造成的间接成本更大的关注。送医类伤害的间接成本数据几乎跟直接成本的类似。即便间接成本和直接成本在数额上非常相似,当工人受伤被当做责任案件对待时,间接成本就会呈指数形式激增。当有多起受伤时,还有非常大的金钱数额要累加到巨大的款项中。有了这样的理解,来看看一个简单的例子,如果试图获得3%的盈利目标,一次伤害可以对一个公司的产量或底线(基本运营成本)造成多大的影响:

If a company acquired a $50,000 loss due to injury, illness or damage and still tries to make a 3 percent profit, the company theoretically must increase sales of services by an additional $1,667,000. [Example taken from Electronic Library of Construction Occupational Safety and Health (eLCOSH)]

如果一家公司因为受伤、疾病或损伤而需承担50000美元的损失,但仍试图创造3%的利润,则该公司理论上必须额外增加高达1667000美元的销售服务。[取自建筑业安全与健康电子数据库(eLCOSH)]

Even with this data there is still unwillingness to embrace safety and health programs fully within some organizations. This reluctance is mainly due to fear and the ability to maintain adequate safety personnel.

即使是有这样的数据,在一些公司机构中仍然会有不愿意完全接受安全和健康计划的现象。这种不情愿主要是由于害怕,同时缺少维持一个适当的安全人员队伍的能力。

Studies show that there is direct correlation between lower accident rate reduction and increased safety investment, and increases safety investment in relation to increased profits. As noted earlier, the implementation of a safety and health program will not have a huge financial impact at the beginning stages of implementation, but given time, it will certainly pay for itself and make the company profit in years to come.

研究表明,降低事故发生率和增加安全投资之间是正相关的,而增加安全投资又与利润的增加有关。如前所述,安全和健康计划的实施将不会在实施的初期阶段造成巨大的财政影响,但只要给予一定的时间,它将肯定能为方案本身买单的,而且会让公司在未来的年头里赢利。

Relationship between construction safety and quality performance
施工安全与质量绩效的关系

It is well established that the project cost, quality, safety, and duration are the four critical elements that contribute to project success. Past literature has established theoretical relationships between construction safety and quality on the basis of opinions of industry experts. This is the first empirical inquiry into the relationship between safety and quality, testing the null hypothesis that there is no statistical relationship among quality performance indicators and safety performance indicators. To test this hypothesis, empirical data were collected from 32 building construction projects, ranging in scope from \$50,000 to \$300 million dollars. Several quality metrics (e.g., cost of rework per \$1M project scope and rate of rework per 200,000 worker-hours) were used as predictor variables and first aid and Occupational Safety and Health Administration (OSHA) recordable injury rates were used as response variables. Linear regressions among the predictor and response variables showed that there are two statistically significant relationships: the OSHA recordable injury rate is positively correlated to rework ($r^2=0.968$; p-value=0.032) and the first aid rate is positively correlated to number of defects ($r^2=0.548$; p-value=0.009).

项目的成本、质量、安全和工期是影响项目成功的四项关键因素，这一点得到了很好的确认。过去的文献已经给出了以行业专家意见为基础的施工安全和质量之间的理论关系。这是对安全和质量之间关系的首次实证调查，检验了质量绩效指标和安全绩效指标之间没有统计关系的零假设。为了检验这一假设，实证数据是从32个建筑施工项目中收集的，范围涵盖从5万美元到3亿美元的项目。几种质量度量（例如，每100万美元项目份额的返工成本和每20万工时的返工率）被用来作为预测变量，而急救率和职业安全与健康管理局（OSHA）可记录的伤害率则用作为反应变量。预测变量和反应变量之间的线性回归表明存在两个显著统计关系：职业安全与健康管理局可记录的伤害率与返工率呈正相关关系（$r^2=0.968$；p值=0.032），而急救率和缺陷数量呈正相关关系（$r^2=0.548$；p值=0.009）。

To understand why these relationships exist and to identify specific strategies that promote both safety and quality, open-ended interviews were conducted with project managers. These individuals indicated that the most compelling reason for the strong positive correlation between rework and injuries is the fact that rework involves demolition, schedule pressure, and unstable work processes. They also noted that devoting resources to preplanning, allowing the necessary time to complete tasks correctly the first time, encouraging leadership at the workface, and encouraging workers to take pride in their work are all strategies that promote both safety and quality.

为了弄清楚为什么存在这些关系，并确定能促进安全和质量的具体对策，和项目经理进行了开放式访谈。这些人表示，返工和受伤之间存在很强的正相关关系最令人信服的原因在于，因为返工会涉及拆卸、工期压力和不稳定的工作流程。他们还指出，将资源投入到预先计划中、给予第一次准确地完成任务所必要的时间、鼓励施工现场的带头作用、鼓励员工对他们的工作感到自豪，这些都是能促进安全和质量的策略。

【Important sentences】

1. In many situations, employers worry about safety and health only as it relates to being regulatory compliant at a minimum with governing agencies.

    only as it 只是因为；relates to 有关，涉及；at a minimum 最低的。

2. But if these companies fully embrace the concepts of implementation and make safety and health a core value, the potential return on investment relating to injury prevention, along with the creation of a safe work environment can be huge, monetarily and culturally.

    fully embrace 完全接纳；make safety and health a core value 把安全与健康作为一种核心价值，core value 表核心价值；relating to … 与……相关；along with 伴随着。

Chapter 7
Disasters in Geotechnical Engineering

岩土工程灾害

7.1 Definition of engineering disasters 工程灾害定义

【Text】

In the field of engineering, the importance of safety is emphasized. Learning from past engineering failures and infamous disasters such as the Challenger explosion brings the sense of reality to what can happen when appropriate safety precautions are not taken. Safety tests such as tensile testing, finite element analysis (FEA), and failure theories help provide information to design engineers about what maximum forces and stresses can be applied to a certain region of a design. These precautionary measures help prevent failures due to overloading and deformation.

Shortcuts in engineering design can lead to engineering disasters. Engineering is the science and technology used to meet the needs and demands of society. These demands include buildings, aircraft, vessels, and computer software. In order to meet social demands, the creation of newer technology and infrastructure must be met efficiently and cost-effectively. To accomplish this, managers and engineers have to have a mutual approach to the specified demand at hand. This can lead to shortcuts in engineering design to reduce costs of construction and fabrication. Occasionally, these shortcuts can lead to unexpected design failures.

Failure occurs when a structure or device has been used past the limits of design that inhibits proper function. If a structure is designed to only support a certain amount of stress, strain, or loading and the user applies greater amounts, the structure will begin to deform and eventually fail. Several factors contribute to failure including a flawed design, improper use, financial costs, and miscommunication.

【Key words】

engineering failure 工程失效，工程事故
failure theory 破坏理论，失效理论
infrastructure n. 基础设施；基础建设
cost-effectively 划算地，物有所值地

【Translation】

In the field of engineering, the importance of safety is emphasized. Learning from past engineering failures and infamous disasters such as the Challenger explosion brings the sense of reality to what can happen when appropriate safety precautions are not taken. Safety tests such as tensile testing, finite element analysis (FEA), and failure theories help provide information to design engineers about what maximum forces and stresses can be applied to a certain region of a design. These precautionary measures help prevent failures due to overloading and deformation.

在工程领域，要强调安全的重要性。从过去的工程事故以及向挑战号发生爆炸这类著名的灾难进行学习能带给我们这种真实的感觉，即当不采取合适的安全措施时可能会发生什么。如拉伸试验、有限元分析（FEA）和失效理论的这类安全测试，有助于为设计工程师们提供关于设计构建中某一区域能够施加的最大的作用力和应力的信息。这些预防措施能帮助防止因超载和变形而发生的破坏。

Shortcuts in engineering design can lead to engineering disasters. Engineering is the science and technology used to meet the needs and demands of society. These demands include buildings, aircraft, vessels, and computer software. In order **to meet social demands**, the creation of newer technology and infrastructure must be met **efficiently and cost-effectively**. To accomplish this, managers and engineers have to have a mutual approach to the specified demand at hand. This can lead to shortcuts in engineering design to reduce costs of construction and fabrication. **Occasionally**, these **shortcuts** can lead to **unexpected** design failures.

工程设计中的捷径会导致工程灾害。工程是满足社会需求和要求的科学技术。这些需求包括建筑物、飞机、船只和计算机软件。为了满足社会的要求，新技术和基础设施的创新必须满足高效化和成本效益化。要做到这一点，管理人员和工程师们必须对手头上指定的需求有一个互通的方法。这会导致工程设计中为了降低建设和制造成本而走捷径。有时候，这些捷径可能导致意料之外的设计缺陷。

Failure occurs when a structure or device has been used past the limits of design that inhibits proper function. If a structure is designed to only support a certain amount of stress, strain, or loading and the user applies greater amounts, the structure will begin to deform and eventually fail. Several factors contribute to failure including a flawed design, improper use, financial costs, and miscommunication.

当一个结构或设备使用超过了设计所规定的限制时，即超过了正常功能范围时，破坏就会发生。如果某个结构设计只能承担一定量的应力、应变或荷载，而使用者却施加了更大的量，那么结构将开始发生变形并有可能失效。会导致发生失效的几项因素包括有缺陷的设计、不当使用、金融成本和误解。

【Important sentences】

1. In order to meet social demands, the creation of newer technology and infrastructure must be met efficiently and cost-effectively.
   meet sb's demands 满足某人的需求；efficiently 高效化；cost-effectively 成本效益化（高性价比）。

2. Occasionally, these shortcuts can lead to unexpected design failures.
   Occasionally, 副词, occasional 的派生词; shortcut 捷径, 可指工程（设计）上的偷工减料; unexpected 过去分词作定语, 意料之外的。

7.2 Basic causes to engineering disasters
造成工程灾害的基本原因

【Text】

Failure due to static loading

Static loading is when a force is applied slowly to an object or structure. Static load tests such as tensile testing, bending tests, and torsion tests help determine the maximum loads that a design can withstand without permanent deformation or failure. Tensile testing is common when calculating a stress-strain curve which can determine the yield strength and ultimate strength of a specific test specimen. Stress-strain curve shows typical yield behavior for ductile metals, as shown in Fig. 7.1. Stress (σ) is shown as a function of strain (ε). Stress and strain are correlated through Young's Modulus

$$\sigma = E\varepsilon \qquad (7.1)$$

Where E is the slope of the linear section of the plot.

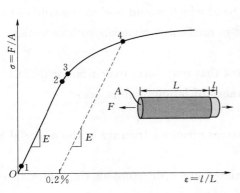

Fig. 7.1　Stress-strain curve for ductile metals
图 7.1　延性金属应力-应变曲线

Fig. 7.2　Tensile test on a composite specimen
图 7.2　复合材料试样的拉伸试验

The specimen is stretched slowly in tension until it breaks, while the load and the distance across the gage length are continuously monitored. A sample subjected to a tensile test can typically withstand stresses higher than its yield stress without breaking. At a certain point, however, the sample will break into two pieces. This happens because the microscopic cracks that resulted from yielding will spread to large scales. The stress at the point of complete breakage is called a material's ultimate tensile strength. The result is a stress-strain curve of the material's behavior under static loading. Through this tensile testing, the yield strength is found at the point where the material begins to yield more readily to the applied stress, and its rate of deformation increases. Tensile test on a composite specimen is shown in Fig. 7. 2.

Failure due to fatigue

When a material undergoes permanent deformation from exposure to radical temperatures or constant loading, the functionality of the material can become impaired. This time-dependent plastic distortion of material is known as creep. Stress and temperature are both major factors of the rate of creep. In order for a design to be considered safe, the deformation due to creep must be much less than the strain at which failure occurs. Once the static loading causes the specimen to surpass this point the specimen will begin permanent, or plastic, deformation.

In mechanical design, most failures are due to time-varying, or dynamic, loads that are applied to a system. This phenomena is known as fatigue failure. Fatigue is known as the weakness in a material due to variations of stress that are repeatedly applied to said material. For example, when stretching a rubber band to a certain length without breaking it (i. e. not surpassing the yield stress of the rubber band) the rubber band will return to its original form after release; however, repeatedly stretching the rubber band with the same amount of force thousands of times would create micro-cracks in the band which would lead to the rubber band being snapped. The same principle is applied to mechanical materials such as metals.

Fatigue failure always begins at a crack that may form over time or due to the manufacturing process used. The three stages of fatigue failure are:

- Crack initiation—when repeated stress creates a fracture in the material being used
- Crack propagation—when the initiated crack develops in the material to a larger scale due to tensile stress
- Sudden fracture failure—caused by unstable crack growth to the point where the material will fail

Note that fatigue does not imply that the strength of the material is lessened after failure. This notion was originally referred to a material becoming "tired" after cyclic loading.

Failure due to miscommunication

Engineering is a precise discipline and in order to be precise, communication among project developers is pertinent for a successful product. There are several forms of miscommunication that can lead to a flawed design in a system. There are various fields of engineering that have to intercommunicate when working toward a mutual goal. These fields include civil, electrical, mechanical, industrial, chemical, biological, and environmental engineering. When creating a modern automobile, electrical engineers, mechanical engineers, and environmental engineers are required to work together to produce a fuel-efficient, durable product for consumers. If engineers do not adequately communicate among one another, a potential design could have flaws and be unsafe for consumer purchase. Engineering disasters can be a result of such miscommunication.

【Key words】

gage n. 厚度；直径；测量仪表；规格　　vt. 计量；度量；估计；判断
microscopic adj. 显微镜的；用显微镜可看见的；微小的
breakage n. 破坏，破损，破损量
fatigue n. 疲劳，疲乏
functionality n. 功能；功能性；设计目的
time-dependent 依时性，与时间有关的
plastic distortion 塑性变形
creep n. & vi. 爬行，蠕动
time-varying adj. 时变的，随时间变化的
snap n. 突然折断　　vi. 啪啪作响，（目光）闪耀
pertinent adj. 有关的，相干的；恰当的
intercommunicate v. 互相联络，互通消息

【Translation】

Failure due to static loading　　静态加载导致的失效

Static loading is when a force is applied slowly to an object or structure. Static load tests such as tensile testing, bending tests, and torsion tests help determine the maximum loads that a design can withstand without permanent deformation or failure. Tensile testing is common when calculating a stress-strain curve which can determine the yield strength and ultimate strength of a specific test specimen. Stress-strain curve shows typical yield behavior for ductile metals, as shown in

Fig. 7.1. Stress (σ) is shown as a function of strain (ε). Stress and strain are correlated through Young's Modulus

$$\sigma = E\varepsilon \tag{7.1}$$

Where E is the slope of the linear section of the plot.

静态加载是指力被慢慢地施加到一个物体或结构上。如拉伸试验、弯曲试验和扭转试验这一类的静态荷载试验，能帮助确定设计在不出现永久变形或破坏时可承担的最大荷载。在计算应力-应变曲线时常用拉伸试验，它可以确定一个特定试样的屈服强度和极限强度。显示延性金属典型屈服性能的应力-应变曲线，如图7.1所示。其中，应力（σ）表示成应变（ε）的函数，应力和应变通过弹性模量产生联系：

$$\sigma = E\varepsilon \tag{7.1}$$

其中 E 是所绘制的曲线线性段的斜率。

The specimen is stretched slowly in tension until it breaks, while the load and the distance across the gage length are continuously monitored. A sample **subjected to a tensile test** can typically withstand stresses **higher than its yield stress without breaking**. At a certain point, however, the sample will break into two pieces. This happens because the microscopic cracks that resulted from yielding will spread to large scales. The stress at the point of complete breakage is called a material's ultimate tensile strength. The result is a stress-strain curve of the material's behavior under static loading. Through this tensile testing, the yield strength is found at the point where the material begins to yield more readily to the applied stress, and its rate of deformation increases. Tensile test on a composite specimen is shown in Fig. 7.2.

拉伸试验中试样被缓慢地拉伸直到发生断裂。经受拉伸试验的试样通常可以承受高于其屈服应力而不断裂的应力。然而，到了某一点试样将断成两块。这种情况的发生是因为受屈服影响产生的细微裂缝将扩展到一个大的范围。对应完全断裂点的应力称为材料的极限抗拉强度。图中结果代表在静载荷作用下的应力-应变曲线。通过这种拉伸试验，可以找到材料开始向所施加的应力产生屈服的位置点，而且它的形变速率也增加了。复合材料试样的拉伸试验如图7.2所示。

Failure due to fatigue 疲劳导致的失效

When a material undergoes permanent deformation from exposure to radical temperatures or constant loading, the functionality of the material can become impaired. **This time-dependent plastic distortion** of material **is known as** creep. Stress and temperature are both major factors of the rate of creep. **In order for a design to be considered safe**, the deformation due to creep must be much less than the strain **at which** failure occurs. Once the static loading causes the specimen to surpass this point the specimen will begin permanent, or plastic, deformation.

当一种材料由于受极端温度或恒定加载而产生永久变形时,材料的性能可能受损。材料的这种与时间有关的塑性变形被称为蠕变。应力和温度两者都是蠕变速率的主要影响因素。为了使一个设计被认为是安全的,由蠕变导致的变形必须要比发生失效时的应变小得多。一旦静态荷载使试样超过这一位置点,试样将开始发生永久性或塑性变形。

In mechanical design, most failures are due to time-varying, or dynamic, loads that are applied to a system. This phenomena is known as fatigue failure. Fatigue is known as the weakness in a material due to variations of stress that are repeatedly applied to said material. For example, when stretching a rubber band to a certain length without breaking it (i.e. not surpassing the yield stress of the rubber band) the rubber band will return to its original form after release; however, repeatedly stretching the rubber band with the same amount of force thousands of times would create micro-cracks in the band which would lead to the rubber band being snapped. The same principle is applied to mechanical materials such as metals.

在机械设计中,大多数的失效是由于系统上施加了随时间改变的或动态的荷载。这种现象被称为疲劳失效。疲劳被认为是一种材料由于受反复作用于所述材料上的应力变化而产生的弱化。例如,当一条橡皮筋拉伸到一定长度而不破坏它(即不超过橡皮筋的屈服应力),橡皮筋可以在放松后恢复到原来的形状;然而,用同样大小的力数以千计地反复拉伸橡皮筋,就会在橡皮筋内部产生细微的裂缝,并可能导致橡皮筋的断裂。同样的原理也适用于机械材料,比如金属。

Fatigue failure always begins at a crack that may form over time or due to the manufacturing process used. The three stages of fatigue failure are:

疲劳失效总是始于因为时间因素或是所使用的加工流程原因而形成的裂缝。疲劳失效的三个阶段是:

- Crack initiation—when repeated stress creates a fracture in the material being used.
  裂纹开裂——交变应力在所使用的材料中形成一个裂缝
- Crack propagation—when the initiated crack develops in the material to a larger scale due to tensile stress
  裂纹扩展——由于拉应力引起初始裂缝在材料中扩展变得更大
- Sudden fracture failure—caused by unstable crack growth to the point

where the material will fail

突然断裂失效——由不稳定的裂缝扩展到材料发生破坏的点所致

Note that fatigue does not imply that the strength of the material is lessened after failure. This notion was originally referred to a material becoming "tired" after cyclic loading.

注意疲劳并不意味着材料的强度在失效后会减少。这个概念最初被称为循环加载后材料"变累"。

Failure due to miscommunication 理解错误导致的失效

Engineering is a precise discipline and in order to be precise, communication among project developers is pertinent for a successful product. There are several forms of miscommunication that can lead to a flawed design in a system. There are various fields of engineering that have to intercommunicate when working toward a mutual goal. These fields include civil, electrical, mechanical, industrial, chemical, biological, and environmental engineering. When creating a modern automobile, electrical engineers, mechanical engineers, and environmental engineers are required to work together to produce a fuel-efficient, durable product for consumers. If engineers do not adequately communicate among one another, a potential design could have flaws and be unsafe for consumer purchase. Engineering disasters can be a result of such miscommunication.

工程是一门严谨的学科，而且为了准确，为使产品成功，项目开发人员之间的沟通是必不可少的。存在几种形式的理解错误，它们能导致系统中有缺陷的设计。当大家为一个共同的目标工作时，各种工程领域必须相互沟通交流。这些领域包括土木、电气、机械、工业、化工、生物和环境工程等领域。在创造一辆现代的汽车时，电气工程师、机械工程师和环境工程师需要共同工作，从而为消费者生产出一个省油、耐用的产品。如果工程师之间没有充分的沟通，设计中就可能会有潜在的缺陷，并且对客户购买造成不安全。工程灾害可以是这种理解错误导致的结果。

【Important sentences】

1. A sample subjected to a tensile test can typically withstand stresses higher than its yield stress without breaking.

   subjected to a tensile test 作 A sample 的定语，表"承受拉伸试验的"；higher than its yield stress without breaking 作 stresses 的后置定语，表"高于其屈服应力而不断裂的"。

2. This time-dependent plastic distortion of material is known as creep.

   Time-dependent plastic distortion 与时间有关的塑性变形；be known as，固定短语，"被称为"。

3. In order for a design to be considered safe, the deformation due to creep must be

much less than the strain at which failure occurs.

In order for ... to be considered safe 为了使……被认为是安全的；at which 引导位置定语。

7.3 Geotechnical engineering disasters 岩土工程灾害

【Text】

Landslides

As catastrophic events, landslides can cause human injury, loss of life and economic devastation, and destroy construction works and cultural and natural heritage.

A landslide, also known as a landslip, is a geological phenomenon that includes a wide range of ground movements, such as rock falls, deep failure of slopes, and shallow debris flows. Landslides can occur in offshore, coastal and onshore environments. Although the action of gravity is the primary driving force for a landslide to occur, there are other contributing factors affecting the original slope stability. Typically, pre-conditional factors build up specific sub-surface conditions that make the area/slope prone to failure, whereas the actual landslide often requires a trigger before being released.

1. Rock fall

A rock fall or rock-fall refers to quantities of rock falling freely from a cliff face. "A rock fall is a fragment of rock (a block) detached by sliding, toppling, or falling, that falls along a vertical or sub-vertical cliff, proceeds down slope by bouncing and flying along ballistic trajectories or by rolling on talus or debris slopes" (Varnes, 1978). Alternatively, a "rock fall is the natural downward motion of a detached block or series of blocks with a small volume involving free falling, bouncing, rolling, and sliding".

2. Rock slope failure

Landslides from massive rock slope failure (MRSF) are a major geological hazard in many parts of the world. Hazard assessment is made difficult by a variety of complex initial failure processes and unpredictable post-failure behaviour, which includes transformation of movement mechanism, substantial changes in volume, and changes in the characteristics of the moving mass. Initial failure mechanisms are strongly influenced by geology and topography. Massive rock slope failure includes rockslides, rock avalanches, catastrophic spreads and rockfalls. Catastrophic debris flows can also be triggered by massive rock slope failure. Volcanoes are particularly prone to massive rock slope failure and can experience very large scale sector collapse or much smaller partial collapse. Both these types of failures may be transformed into lahars which can travel over 100 km from their source. MRSF deposits give insight into fragmentation and emplacement processes. Slow mountain slope deformation presents problems in interpretation of origin and movement mechanism. The identification of thresholds for the catastrophic failure of a slow moving rock slope is a key question in hazard assessment. Advances have been made in the analysis and modeling of initial failure and post-failure behaviour. However, these studies have been retrodictive in nature and their true predictive potential for hazard assessment remains uncertain yet promising. These processes, which can be instantaneous or delayed, include the formation and failure of landslide dams and the generation of landslide tsunamis. Both these processes extend potential damage beyond the limits of landslide debris. The occurrence of MRSF forms orderly magnitude and frequency relations which can be characterized by robust power law relationships. MRSF is increasingly recognized as being an important process in landscape evolution which provides an essential context for enhanced hazard assessment. Secondary processes associated with MRSF are an important component of hazard.

3. Debris flow

Debris flows are geological phenomena in which water-laden masses of soil and fragmented rock rush down mountainsides, funnel into stream channels, entrain objects in their paths, and form thick, muddy deposits on valley floors. They generally have bulk densities comparable to those of rock avalanches and other types of landslides (roughly 2,000 kilograms per cubic meter), but owing to widespread sediment liquefaction caused by high pore-fluid pressures, they can flow almost as fluidly as water. Debris flows descending steep channels commonly attain speeds that surpass 10 meters per second, although some large flows can reach speeds that are much greater. Debris flows with volumes ranging up to about 100,000 cubic meters occur frequently in mountainous regions worldwide. The largest prehistoric flows have had volumes exceeding 1 billion cubic meters (i.e., 1 cubic kilometer). As a result of their high sediment concentrations and mobility, debris flows can be very destructive.

Subsidence

Subsidence is the motion of a surface (usually, the earth's surface) as it shifts downward relative to a datum such as sea-level. The opposite of subsidence is uplift, which results in an increase in elevation. Ground subsidence is of concern to geologists, geotechnical engineers and surveyors.

7.3 Geotechnical engineering disasters 岩土工程灾害

1. Dissolution of limestone

Subsidence frequently causes major problems in karst terrains, where dissolution of limestone by fluid flow in the subsurface causes the creation of voids (i.e. caves). If the roof of these voids becomes too weak, it can collapse and the overlying rock and earth will fall into the space, causing subsidence at the surface. This type of subsidence can result in sinkholes which can be many hundreds of meters deep.

2. Mining

Several types of sub-surface mining, and specifically methods which intentionally cause the extracted void to collapse (such as pillar extraction, longwall mining and any metalliferous mining method which uses "caving" such as "block caving" or "sub-level caving") will result in surface subsidence. Mining-induced subsidence is relatively predictable in its magnitude, manifestation and extent, except where a sudden pillar or near-surface underground tunnel collapse occurs (usually very old working). Mining-induced subsidence is nearly always very localized to the surface above the mined area, plus a margin around the outside. The vertical magnitude of the subsidence itself typically does not cause problems, except in the case of drainage (including natural drainage) —rather, it is the associated surface compressive and tensile strains, curvature, tilts and horizontal displacement that are the cause of the worst damage to the natural environment, buildings and infrastructure where mining activity is planned, mining-induced subsidence can be successfully managed if there is co-operation from all of the stakeholders. This is accomplished through a combination of careful mine planning, the taking of preventive measures, and the carrying out of repairs post-mining.

3. Extraction of natural gas

If natural gas is extracted from a natural gas field, the initial pressure [up to 60 MPa (600 bar)] in the field will drop over the years. The gas pressure also supports the soil layers above the field. If the pressure drops, the soil pressure increases and this leads to subsidence at the ground level.

Since exploitation of the Slochteren Netherlands gas field started in the late 1960s the ground level over a 250 km² area has dropped by a current maximum of 30 cm.

4. Earthquake

The Geospatial Information Authority of Japan reported immediate subsidence caused by the 2011 Tōhoku earthquake In Northern Japan, subsidence of 0.50 m (1.64 feet) was observed on the coast of the Pacific Ocean in Miyako, Tōhoku, while Rikuzentakata, Iwate measured 0.84m (2.75 feet). In the south at Sōma, Fukushima, 0.29m (0.95 feet) was observed. The maximum amount of subsidence was 1.2m (3.93 feet), coupled with horizontal diastrophism of up to 5.3m (17.3 feet) on the Oshika Peninsula in Miyagi Prefecture.

5. Groundwater-related subsidence

The habitation of lowlands, such as coastal or delta plains, requires drainage. The resulting aeration of the soil leads to the oxidation of its organic components, such as peat, and this decomposition process may cause significant land subsidence. This applies especially when ground water levels are periodically adapted to subsidence, in order to maintain desired unsaturated zone depths, exposing more and more peat to oxygen. In addition to this, drained soils consolidate as a result of increased effective stress. In this way, land subsidence has the potential of becoming self-perpetuating, having rates up to 5cm/a. Water management used to be tuned primarily to factors such as crop optimisation but, to varying extents, avoiding subsidence has come to be taken into account as well.

6. Faulting induced

When differential stresses exist in the Earth, these can be accommodated either by geological faulting in the brittle crust, or by ductile flow in the hotter and more fluid mantle. Where faults occur, absolute subsidence may occur in the hanging wall of normal faults. In reverse, or thrust, faults, relative subsidence may be measured in the footwall.

7. Isostatic subsidence

The crust floats buoyantly in the plastic as the Slochteren nosphere, with a ratio of mass below the "surface" in proportion to its own density and the density of the as the nosphere. If mass is added to a local area of the crust (e.g. through deposition), the crust subsides to compensate and maintain isostatic balance.

The opposite effect to isostatic subsidence is known as isostatic rebound—the action of the crust returning (sometimes over periods of thousands of years) to a state of isostasy, such as after the melting of large ice sheets or the drying-up of large lakes after the last ice age. Lake Bonneville is a famous example of isostatic rebound. Due to the weight of the water once held in the lake, the Earth's crust subsided nearly 200 feet (61m) to maintain equilibrium, when the lake dried up, the crust rebounded. Today at modern Lake Bonneville, the center of the former lake is about 200 feet (61m) higher than the former lake edges.

8. Seasonal effects

Many soils contain significant proportions of clay which because of the very small particle size are affected by changes in soil moisture content. Seasonal drying of the soil results in a reduction in soil volume and a lowering of the soil surface. If building foundations are above the level to which the seasonal drying reaches they will move and this can result in damage to the building in the form of tapering cracks. Trees and other vegetation can have a significant local effect on seasonal drying of soils. Over a number of years a cumulative drying occurs as the tree grows, this can lead to the opposite of subsidence, known as heave or swelling of the soil, when the tree declines or is felled. As the cumulative moisture deficit is reversed, over a period which can last as many as 25 years, the surface level around the tree will rise and expand laterally. This is often more damaging to buildings unless the foundations have been strengthened or designed to cope with the effect.

【Key words】

landslip *n.* 山崩；地滑；崩塌；塌方
trigger *vt.* 引发，触发
cliff face 悬崖壁
detach *vt.* 分离，拆开；派遣
topple *vi.* 倾倒；摇摇欲坠
ballistic trajectory 弹道轨迹
talus（=debris slope）*n.* 斜面；塌砾；岩堆（岩屑坡）
post-failure 破坏后区
lahar *n.* ［爪哇语］火山泥流
fragmentation *n.* 分裂，破碎
emplacement *n.* 放列动作；炮兵掩体
threshold *n.* 门槛，入口，开始；［物理学］临界值
landslide dam 堰塞湖
power law relationship 幂律关系
landscape evolution 景观变化，景观变迁，景观演化
secondary process 次级过程
water-laden 承水的
stream channel 河道河槽
entrain *v.* 使乘火车；拖；产生
valley floor 谷底，河谷
bulk density 堆积密度
rock avalanche 岩崩
pore-fluid pressure 孔隙水压力
mountainous *adj.* 多山的；巨大的；山一般的
sediment concentration 含沙量；沉积物浓度

datum n. 数据，资料；[测]基点，基线，基面
uplift vt. 举起；振作　vi. 上升，升起
elevation n. 高处，高地，高度，海拔
karst terrain　喀斯特地貌，岩溶地貌
fluid flow　流体流动
void adj. 空的，空虚的　n. 太空，宇宙空间；空位，空隙
sinkhole n. 污水池
manifestation n. 表示，显示；示威
curvature n. 弯曲；弯曲部分；曲率
tilt vt. 使倾斜；拿枪扎；抨击
stakeholder n. 股东；利益相关者
Slochteren　（荷兰）斯洛赫特伦
Miyako, Tōhoku　东北町宫古市
Rikuzentakata, Iwate　岩手町陆前高田市
Sōma, Fukushima　福岛町相马市
ft（=feet）n. 英尺（=12in 或 30.48cm）
horizontal diastrophism　地壳水平运动
Oshika Peninsula　牡鹿半岛
Miyagi Prefecture　宫城县
lowland n. 低地　adj. 低地的
aeration n. 通风；换气；松砂；增氧
oxidation n. 氧化
peat n. 泥煤，泥炭
decomposition process　分解过程
unsaturated zone depth　包气带厚度
self-perpetuating　能使自身永久存在的；自我延续
water management　水管理；用水管理
optimization n. 最佳化，最优化；优选法；优化组合
brittle adj. 易碎的；难以相处的，尖刻暴躁的
crust n. 面包皮；外壳；硬外皮；地壳
ductile flow　黏性流变；塑性流动
fluid mantle　地幔流体
thrust vt. & vi. 猛推；逼迫；强行推入；延伸　n. 刺；推力
isostatic adj. 均衡说的
buoyant adj. 轻快的；活泼的，开朗的；有浮力的，易浮的
rebound vt. 使弹回；使回升；使回响　n. 弹回，跳回
isostasy n. 地壳均衡说
Lake Bonneville　邦纳维尔湖（史前时期存在于北美洲的巨大湖泊）
soil moisture content　土壤含水量
tapering adj. 尖端细的

cumulative *adj.* 累积的；渐增的；追加的
heave *vi.* 起伏；（山丘等）隆起；拖；气喘 *n.* 举起；波动；隆起
swelling *n.* 肿胀；膨胀；增大 *v.* 肿胀
fell *vt.* 砍倒，打倒

【Translation】

Landslides 山体滑坡

As catastrophic events, landslides can cause human injury, loss of life and economic devastation, and destroy construction works and cultural and natural heritage.

作为灾难性的事件，山体滑坡会造成人身伤害、生命损失和经济灾难，也会摧毁施工工程和文化与自然遗产。

A landslide, also known as a landslip, is a geological phenomenon **that** includes a wide range of ground movements, such as rock falls, deep failure of slopes, and shallow debris flows. Landslides can occur in offshore, coastal and onshore environments. Although the action of gravity is the primary driving force for a landslide to occur, there are other contributing factors affecting the original slope stability. **Typically, pre-conditional factors build up** specific sub-surface conditions **that** make the area/slope **prone to** failure, **whereas** the actual landslide often requires a trigger before being released.

山体滑坡，也被称为山泥倾泻，是一种地质现象，它包括分布极广的岩土运动，比如岩石崩落、边坡深层失效破坏和浅层泥石流。山体滑坡可以发生在离岸、海岸和内陆等环境。虽然重力作用是山体滑坡发生的主要驱动力，但也存在一些影响原始边坡稳定性的其他作用因素。通常情况下，先天条件因素形成了特定的地下条件，不利的条件会使该区域或边坡更容易出现失效，然而实际的山体滑坡发生滑坡前往往需要某个触发因素。

1. Rock fall 岩石崩落

A rock fall or rock-fall refers to quantities of rock falling freely from a cliff face. "A rock fall is a fragment of rock (a block) detached by sliding, toppling, or falling, that falls along a vertical or sub-vertical cliff, proceeds down slope by bouncing and flying along ballistic trajectories or by rolling on talus or debris slopes" (Varnes, 1978). Alternatively, a "rock fall is the natural downward motion of a detached block or series of blocks with a small volume involving free falling, bouncing, rolling, and sliding".

岩石崩落或岩崩是指大量岩石从一个悬崖面上自由崩落。"岩石崩落是指因为滑动、倾倒或下坠而分离的岩石碎片（岩块）沿着垂直或近似垂直的悬崖下落，并沿着弹道轨迹通过弹跳和飞行方式，或者在岩屑坡上滚动的方式落向坡下"（Varnes，1978）。或者说，"岩石崩落是指分离的岩块或小体积的岩块群的自然向下运动，涉及自由下落、弹跳、滚动和滑动"。

2. Rock slope failure 岩石边坡失效

Landslides from massive rock slope failure (MRSF) are a major geological hazard in many parts of the world. **Hazard assessment is made difficult by** a variety of complex initial failure processes and unpredictable post-failure behaviour, **which** includes transformation of movement mechanism, substantial changes in volume, and changes in the characteristics of the moving mass. Initial failure mechanisms are strongly influenced by geology and topography. Massive rock slope failure includes rockslides, rock avalanches, catastrophic spreads and rockfalls. Catastrophic debris flows can also be triggered by massive rock slope failure. Volcanoes are particularly prone to massive rock slope failure and can experience very large scale sector collapse or much smaller partial collapse. Both these types of failures may be transformed into lahars which can travel over 100km from their source. MRSF deposits give insight into fragmentation and emplacement processes. Slow mountain slope deformation presents problems in interpretation of origin and movement mechanism. The identification of thresholds for the catastrophic failure of a slow moving rock slope is a key question in hazard assessment. Advances have been made in the analysis and modeling of initial failure and post-failure behaviour. However, these studies have been retrodictive in nature and their true predictive potential for hazard assessment remains uncertain yet promising. These processes, which can be instantaneous or delayed, include the formation and failure of landslide dams and the generation of landslide tsunamis. Both these processes extend potential damage beyond the limits of landslide debris. The occurrence of MRSF forms orderly magnitude and frequency relations which can be characterized by robust power law relationships. MRSF is increasingly recognized as being an important process in landscape evolution which provides an essential context for enhanced hazard assessment. Secondary processes associated with MRSF are an important component of hazard.

大规模岩石边坡失稳所引起的山体滑坡是在世界许多地区存在的一类严重的地质灾害。灾害评估并非易事，它由各种复杂的初始失效过程和不可预知的失效后行为得到，这些行为包括运动机理的转换、体积的实质性变化以及运动体特征的改变。地质和地形条件对初始失效机制有强烈的影响。大规模岩石边坡失稳包括滑坡、岩崩、灾难性扩散和落石。大规模岩石边坡失稳也可以触发灾难性的泥石流。火山群特别容易发生大规模岩石边坡失稳，可以表现为非常大型的山体崩塌或更小的局部坍塌。这两种类型的失效都可能演变成火山泥流，它可以从源头开始穿行

100多km之遥。大规模岩石边坡失稳形成的堆积物提供了一个深入了解块度分布和堆砌过程的机会。缓慢的山体边坡变形能代表用于解释成因及运动机理的问题。缓慢移动的岩质边坡突变失稳的触发点的确定是灾害评估中的关键。在初始失效和失效后的行为模拟与分析方面已经取得了不少进展。然而，这些研究在特性上一直是倒推式的，对灾害评估的真实的可预测潜力虽然不确定，但还是有前途的。这些过程可以是同步发生的或滞后的，包括滑坡坝体的形成和失效，以及山崩海啸的产生等。这两种过程都超出了滑坡泥石流可能造成的潜在伤害的极限。大规模岩石边坡失稳的发生率依次形成了幅度和频率关系，它可以通过强大的幂率关系来表现。大规模岩石边坡失稳越来越被公认为是坡体形成的一个重要过程，这能为加强灾害评估提供重要的参考。与大规模岩石边坡失稳相关的次级过程也是灾害的一个重要组成部分。

3. Debris flow 泥石流

Debris flows are geological phenomena in which water-laden masses of soil and fragmented rock rush down mountainsides, funnel into stream channels, entrain objects in their paths, and form thick, muddy deposits on valley floors. They generally have bulk densities comparable to those of rock avalanches and other types of landslides (roughly 2,000 kilograms per cubic meter), but owing to widespread sediment liquefaction caused by high pore-fluid pressures, they can flow almost as fluidly as water. Debris flows descending steep channels commonly attain speeds that surpass 10 meters per second, although some large flows can reach speeds that are much greater. Debris flows with volumes ranging up to about 100,000 cubic meters occur frequently in mountainous regions worldwide. The largest prehistoric flows have had volumes exceeding 1 billion cubic meters (i.e., 1 cubic kilometer). As a result of their high sediment concentrations and mobility, debris flows can be very destructive.

泥石流是一种地质现象，泥石流发生时，土壤和岩石碎片的富水体冲下山坡，汇集成洪流，卷走所经之处的各类物体，并在谷底形成厚厚的泥质沉积物。一般来说，它们具有的体积密度跟岩石崩滑体以及其他类型的滑坡体的体积密度相当（大约2000kg/m³），但由于高强度的孔隙水压力会导致大范围的沉积液化，他们几乎可以像流水一样流动。泥石流沿陡峭通道下泄时通常能达到超过10m/s的速度，尽管一些大规模的河流可以达到大得多速度。世界各地山区中经常发生的泥石流的体积变化范围可高达100000m³左右。最大的史前泥石流的体积超过了10亿m³（即1km³）。由于泥石流含有大量沉积物而且具有高流动性，因此可以具有非常大的破坏性。

Subsidence 沉降下陷

Subsidence is the motion of a surface (usually, the Earth's surface) as it shifts downward relative to a datum such as sea-level. The opposite of subsidence is uplift, which results in an increase in elevation. Ground subsidence is of concern to geologists, geotechnical engineers and surveyors.

沉降下陷指某表面（通常如地球表面）相对于一个基准面，比如海平面，产生向下的运动。与沉降下陷相反的是隆起，它导致地面升高。地面沉降是地质学家、岩土工程师和测量人员需要关注的。

1. Dissolution of limestone 石灰岩溶解

Subsidence frequently causes major problems in karst terrains, where dissolution of limestone by fluid flow in the subsurface causes the creation of voids (i.e. caves). If the roof of these voids becomes too weak, it can collapse and the overlying rock and earth will fall into the space, causing subsidence at the surface. This type of subsidence can result in sinkholes which can be many hundreds of meters deep.

在喀斯特地区，沉降下陷频繁造成大的问题，因为地下水流对石灰岩溶蚀作用导致空洞（溶洞）的生成。如果这些空洞的顶部变得太弱，就会崩塌，此时上覆岩石和土体将掉入下部空间，从而引起地面沉降。这种类型的沉降下陷能引起数百米深的塌陷坑。

2. Mining 采矿

Several types of sub-surface mining, and specifically methods which intentionally cause the extracted void to collapse (such as pillar extraction, longwall mining and any metalliferous mining method which uses "caving" such as "block caving" or "sub-level caving") will result in surface subsidence. Mining-induced subsidence is relatively predictable in its magnitude, manifestation and extent, except where a sudden pillar or near-surface underground tunnel collapse occurs (usually very old working). Mining-induced subsidence is nearly always very localized to the surface above the mined area, plus a margin around the outside. The vertical magnitude of the subsidence itself typically does not cause problems, except in the case of drainage (including natural drainage) —rather, it is the associated surface compressive and tensile strains, curvature, tilts and horizontal displacement that are the cause of the worst damage to the natural environment, buildings and infrastructure where mining activity is planned, mining-induced subsidence can be successfully managed if there is co-operation from all of the stakeholders. This is accomplished through a combination of careful mine planning, the taking of preventive measures, and the carrying out of repairs post-mining.

有几种地下开采类型以及特定的通过故意形成采空区来引发坍塌的采矿方法（如矿柱回采、长壁采煤和采用诸如"崩落开采法""分块崩落开采法"或"分段崩落开采法"的任何金属矿采矿方法),将导致地表沉陷。采矿引起的沉陷在范围、表现形式和程度上是相对可预测的,除了出现矿柱或近地表地下隧道突然发生塌方(通常在老采区)。采矿引起的地表沉降几乎总是刚好位于采空区上方的地表,再加上一定的外部边缘。沉降的垂直幅度本身通常不会造成什么问题,除非是在排水的情况下(包括自然排水)——相反,相关的地面压缩和拉伸应变、弯曲、翘起和水平位移才是对自然环境、计划用于采矿活动的建筑和基础设施等造成最严重损害的原因,而如果可以得到所有利益相关者的合作支持,采矿引起的沉降下陷问题则可以得到成功管控。通过采取细致的采矿规划、采取预防措施并进行采矿作业后期的修复等综合措施就能完成这一目标。

3. Extraction of natural gas 天然气的开采

If natural gas is extracted from a natural gas field, the initial pressure [up to 60MPa (600 bar)] in the field will drop over the years. The gas pressure also supports the soil layers above the field. If the pressure drops, the soil pressure increases and this leads to subsidence at the ground level.

如果天然气采自天然气田,天然气田的初始压力[高达60MPa(即600bar)]将下降数年。天然气的气体压力也支撑着采区上方的土体。如果压力下降,土压力增大,就会导致地平面沉降下陷。

Since exploitation of the Slochteren Netherlands gas field started in the late 1960s the ground level over a 250km² area has dropped by a current maximum of 30cm.

荷兰斯洛赫特伦的天然气田从20世纪60年代后期开始开发利用以来,有超过250km² 的地平面已经出现了到目前为止最大值达30cm的沉降。

4. Earthquake 地震

The Geospatial Information Authority of Japan reported immediate subsidence caused by the 2011 Tōhoku earthquake In Northern Japan, subsidence of 0.50m (1.64 feet) was observed on the coast of the Pacific Ocean in Miyako, Tōhoku, while Rikuzentakata, Iwate measured 0.84m (2.75 feet). In the south at Sōma, Fukushima, 0.29m (0.95 feet) was observed. The maximum amount of subsidence was 1.2m (3.93 feet), coupled with horizontal diastrophism of up to 5.3m (17.3 feet) on the Oshika Peninsula in Miyagi Prefecture.

日本地理空间信息管理局报道称,2011年在日本北部东北町发生的地震引起了直接沉降,在东北町宫古市太平洋海岸上观测到了0.50m(1.64ft)的沉降,而在岩手町陆前高田市观测值则为0.84m(2.75ft)。在福岛町相马市的南部,观测值为0.29m(0.95ft)。在宫城县的牡鹿半岛最大沉降量达到1.2m(3.93ft),伴随高达 5.3m（17.3ft）的水平地壳变形。

5. Groundwater-related subsidence 与地下水有关的沉陷

The habitation of lowlands, such as coastal or delta plains, requires drainage. The resulting aeration of the soil leads to the oxidation of its organic components, such as peat, and this decomposition process may cause significant land subsidence. This applies especially when ground water levels are periodically adapted to subsidence, in order to maintain desired unsaturated zone depths, exposing more and more peat to oxygen. In addition to this, drained soils consolidate as a result of increased effective stress. In this way, land subsidence has the potential of becoming self-perpetuating, having rates up to 5cm/a. Water management used to be tuned primarily to factors such as crop optimisation but, to varying extents, avoiding subsidence has come to be taken into account as well.

在低地居住，如沿海或三角洲平原，是需要排水的。由此产生的土壤气孔会导致其有机组成部分，比如腐殖质的氧化，这种分解过程可以导致明显的地面沉降。这特别适合当地下水位呈周期性地适应沉降的时候，为了保持所需的非饱和带深度，就会暴露更多的腐殖质到氧气环境中。除此之外，由于有效应力的增加而导致的疏干土壤出现固结。通过这种方式，地面沉降就具有自我保持的潜力，具有高达 5cm/a 的速度。水管理曾是作物优化中进行调整的首要因素，但是，在不同区段，避免沉降也已经纳入考虑的范围内。

6. Faulting induced 断裂作用引起

When differential stresses exist in the Earth, these can be accommodated either by geological faulting in the brittle crust, or by ductile flow in the hotter and more fluid mantle. Where faults occur, absolute subsidence may occur in the hanging wall of normal faults. In reverse, or thrust, faults, relative subsidence may be measured in the footwall.

地层中存在有差异的应力，它们可以或者通过脆性地壳中的地质断层运动，或者通过更热和更具流动性的地幔的可延性流动体现出来。当出现断层时，正断层上盘中可能会出现绝对沉降。相反，在逆断层中，可能在下断层下盘中测到相对沉降。

7. Isostatic subsidence 地壳均衡沉降

The crust floats buoyantly in the plastic as the Slochteren nosphere, with a ratio of mass below the "surface" in proportion to its own density and the density of the as the nosphere. If mass is added to a local area of the crust (e.g. through deposition), the crust subsides to compensate and maintain isostatic balance.

地壳漂浮在像斯洛赫特伦软流一样的塑料体上，位于"地表"以下的质量比例与其自身的密度和作为软流的密度成正比关系。如果一些物体加到地壳的局部区域上（例如通过沉积作用），地壳就通过沉降来进行补偿并保持地壳均衡的平衡性。

The opposite effect to isostatic subsidence is known as isostatic rebound—the action of the crust returning (sometimes over periods of thousands of years) to a state of isostasy, such as after the melting of large ice sheets or the drying-up of large lakes after the last ice age. Lake Bonneville is a famous example of isostatic rebound. Due to the weight of the water once held in the lake, the Earth's crust subsided nearly 200 feet (61m) to maintain equilibrium, when the lake dried up, the crust rebounded. Today at modern Lake Bonneville, the center of the former lake is about 200 feet (61m) higher than the former lake edges.

与地壳均衡沉降相反的作用被称地壳均衡反弹——即地壳恢复到一个均衡状态的活动（有时是数千年的周期），比如上次的冰河世纪后的大冰盖的融化或大型湖泊的干枯。邦纳维尔湖是地壳均衡反弹的著名案例。由于湖里曾经蓄水的重量影响，为了保持平衡，地壳下降接近200ft（61m），当湖干涸时，地壳反弹了。今天，在现在的邦纳维尔湖城，原先湖的中心比其边缘高出大约200ft（61m）。

8. Seasonal effects　季节效应

Many soils contain significant proportions of clay which because of the very small particle size are affected by changes in soil moisture content. Seasonal drying of the soil results in a reduction in soil volume and a lowering of the soil surface. If building foundations are above the level to which the seasonal drying reaches they will move and this can result in damage to the building in the form of tapering cracks. Trees and other vegetation can have a significant local effect on seasonal drying of soils. Over a number of years a cumulative drying occurs as the tree grows, this can lead to the opposite of subsidence, known as heave or swelling of the soil, when the tree declines or is felled. As the cumulative moisture deficit is reversed, over a period which can last as many as 25 years, the surface level around the tree will rise and expand laterally. This is often more damaging to buildings unless the foundations have been strengthened or designed to cope with the effect.

许多土壤中含有可观比例的黏土，并因为颗粒尺寸很小而受到土壤湿度改变的影响。土壤的季节性干燥会导致土壤体积的减少，并使土壤表面降低。如果建筑地基位于季节性干燥能达到的水平之上，他们将会发生移动，而这可以导致建筑物出现尖细裂缝形式的损害。树木及其他植被对土壤季节性干燥有明显的局部影响。随着树生长的许多年内会出现累积干燥，当树木衰老或被伐倒的时候，会导致被称为隆起或土壤膨胀的反沉降。随着累积的含水量不足出现反转，在长达25年的时间内，树木周围的地面水平会上升并向侧向扩展。这往往对建筑物更具破坏性，除非地基已经进行了加固，或专门针对这种影响进行了设计。

【Important sentences】

1. A landslide, also known as a landslip, is a geological phenomenon that includes a wide range of ground movements, such as rock falls, deep failure of slopes, and shallow debris flows.

   also known as a landslip 作插入语；that 引导宾语从句。

2. Typically, pre-conditional factors build up specific sub-surface conditions that make the area/slope prone to failure, whereas the actual landslide often requires a trigger before being released.

   Typically 通常情况下；pre-conditional factors 先天条件因素；build up 建立、形成；that 引导宾语从句；prone to 易于；whereas，连接词，表转折。

3. Hazard assessment is made difficult by a variety of complex initial failure processes and unpredictable post-failure behavior, which includes transformation of movement mechanism, substantial changes in volume, and changes in the characteristics of the moving mass.

   Hazard assessment 灾害评估；be made difficult by… 通过……做起来很难；which 引导非限制性定语从句。

7.4 Engineering disasters prevention 工程灾害防治

【Text】

Disaster mitigation and rehabilitation have become some of the most pressing topics for discussion in geotechnical engineering and related professions for many years. Disaster management encompasses diverse topics such as natural disasters (tsunamis, earthquakes, landslides, etc.), mitigation (early warning and prediction systems, hazard mapping, risk analysis, coastal protection works, etc.), rehabilitation and reconstruction (difficult soils and ground treatment, design against earthquakes and other natural disasters, etc.), and many others, including soil dynamics, liquefaction, stability, and environmental protection.

Disaster mitigation

Disaster mitigation measures are those that eliminate or reduce the impacts and risks of hazards through proactive measures taken before an emergency or disaster occurs.

One of the best known examples of investment in disaster mitigation is the Red River Floodway. The building of the Floodway was a joint provincial/federal undertaking to protect the City of Winnipeg and reduce the impact of flooding in the Red River Basin. It cost $60 million to build in the 1960s. Since then, the floodway has been used over 20 times. Its use during the 1997 Red River Flood alone saved an estimated $6 billion. The Floodway was expanded in 2006 as a joint provincial/federal initiative.

All-hazards approach

An all-hazards emergency management approach looks at all potential risks and impacts, natural and human-induced (intentional and non-intentional) to ensure that decisions made to mitigate against one type of risk do not increase our vulnerability to other risks.

Types of disaster mitigation

Disaster mitigation measures may be structural (e.g. flood dikes) or non-structural (e.g. land use zoning). Mitigation activities should incorporate the measurement and assessment of the evolving risk environment. Activities may include the creation of comprehensive, pro-active tools that help decide where to focus funding and efforts in risk reduction.

Other examples of mitigation measures include:

- Hazard mapping
- Adoption and enforcement of land use and zoning practices
- Implementing and enforcing building codes
- Flood plain mapping
- Reinforced city safe rooms
- Burying of electrical cables to prevent ice build-up
- Raising of homes in flood-prone areas
- Disaster mitigation public awareness programs
- Insurance programs

Recovery from disasters

Disasters are inevitable but mostly unpredictable, and they vary in type and magnitude. The best strategy is to have some kind of disaster recovery plan in place, to return to normal after the disaster has struck.

In the event of a large-scale natural disaster where response and recovery costs exceed what individual provinces and territories could reasonably be expected to bear on their own, the central government has to provide necessary assistances including financial support to the provincial and territorial governments at the first time. Assistance is paid to the province or territory—not directly to individuals or communities. The provincial or territorial governments design, develop and deliver disaster financial assistance, determining the amounts and types of assistance that will be provided to those who have experienced losses.

Every business disaster has one or more causes and effects. The causes can be natural or human or mechanical in origin, ranging from events such as a tiny hardware or software component's malfunctioning to universally recognized events such as earthquakes, fire and flood. Effects of disasters range from small interruptions to total business shutdown for days or months, even fatal damage to the business.

The process of preparing a disaster recovery plan begins by identifying these causes and effects, analyzing their likelihood and severity, and ranking them in terms of their business priority. The ultimate results are a formal assessment of risk, a disaster recovery plan that includes all available recovery mechanisms, and a formalized Disaster Recovery Committee that has responsibility for rehearsing, carrying out, and improving the disaster recovery plan.

The disaster recovery plan should: ①identify and classify the threats/risks that may lead to disasters; ②define the resources and processes that ensure business continuity during the disaster; ③define the reconstitution mechanism to get the business back to normal from the disaster recovery state, after the effects of the disaster are mitigated. An effective disaster recovery plan plays its role in all stages of the operations as depicted above, and it is continuously improved by disaster recovery mock drills and feedback capture processes.

The methods and procedures involved in the disaster recovery planning process

(1) The first step in planning recovery from unexpected disasters is to identify the threats or risks that can bring about disasters by doing risk analysis.

(2) When evaluating risks, it is recommended to categorize them into different classes to accurately prioritize them. In general, risks can be classified in the following five categories.: External Risks, Facility Risks, Data Systems Risks, Departmental Risks, Desk-Level Risks.

(3) Once the evaluation of the major risk categories is completed, it is time to score and sort all of them, category by category, in terms of their likelihood and impact. The scoring process can be approached by preparing a score sheet.

(4) Once the disaster risks have been assessed and the decision has been made

to cover the most critical risks, the next step is to determine and list the likely effects of each of the disasters. These specific effects are what will need to be covered by the disaster recovery process.

(5) Once the list of affected entities is prepared and each entity's business criticality and failure tendency is assessed, it is time to analyze various recovery methods available for each entity and determine the best suitable recovery method for each. This step defines the resources employed in recovery and the process of recovery.

(6) Disaster recovery operations and procedures should be governed by a central committee. This committee should have representation from all the different company agencies with a role in the disaster recovery process. The Disaster Recovery Committee creates the disaster recovery plan and maintains it. During a disaster, this committee ensures that there is proper coordination between different agencies and that the recovery processes are executed successfully and in proper sequence.

The different phases of disaster recovery

Disaster recovery happens in the following sequential phases.

(1) Activation Phase. In this phase, the disaster effects are assessed and announced. The activation phase involves:

- Notification procedures
- Damage assessment
- Disaster recovery activation planning

(2) Execution Phase. In this phase, the actual procedures to recover each of the disaster affected entities are executed. Business operations are restored on the recovery system. Recovery operations start just after the disaster recovery plan has been activated, appropriate operations staff have been notified, and appropriate teams have been mobilized. The activities of this phase focus on bringing up the disaster recovery system. Depending on the recovery strategies defined in the plan, these functions could include temporary manual processing, recovery and operation on an alternate system, or relocation and recovery at an alternate site.

(3) Reconstitution Phase. In this phase the original system is restored and execution phase procedures are stopped. In the reconstitution phase, operations are transferred back to the original facility once it is free from the disaster aftereffects,

and execution-phase activities are subsequently shut down. If the original system or facility is unrecoverable, this phase also involves rebuilding. Hence the reconstitution phase may last for a few days to few weeks or even months, depending on the severity of destruction and the site's fitness for restoration. As soon as the facility, whether repaired or replaced, is able to support its normal operations, the services may be moved back. The execution team should continue to be engaged until the restoration and testing are complete.

What information the disaster recovery plan should contain

The outcome of the disaster recovery planning process is the disaster recovery plan document. During an emergency, this document will be the primary source of information for disaster recovery procedures.

The disaster recovery plan document is the only reliable source of information for the disaster recovery during an emergency. It should be very easily readable, with simple and detailed instructions. Some of the contents that need to be in this document are as follows:

Document Information/Purpose/Scope/Assumptions/Exclusions/System Description/Roles and Responsibilities/Contact Details/Activation Procedures/Execution Procedures/Reconstitution Procedures.

How to maintain the disaster recovery plan

The disaster recovery plan document needs to be kept up to date with the current organization environment. A plan that is not updated and tested is as bad as not having a plan at all because during emergencies, the document may be misleading. The following are recommended for maintenance of the plan documentation.

- Periodic mock drills: The disaster recovery plan should be tested from time to time using scheduled mock drills. A drill usually will not affect active operations; however, if it is known that operations will be affected, the drill should be carefully scheduled such that the effect is minimal and is done during a permissible window. These activities should be regarded similarly to regular equipment maintenance activities that require operations downtime. The experience of the mock drill should be updated into the disaster recovery plan document.
- Experience capture: The best testing the document will undergo is when an actual disaster happens, and the lessons learned during the disaster recovery are valuable for improving the plan. Hence the Disaster Recovery Committee should ensure that the experience gets captured as lessons

learned and the document gets updated accordingly.
- Periodic updates: Technologies, systems, and facilities that the plan covers may change over time. It is important that the disaster recovery plan document reflect the current information about the components it covers. For this purpose, the Disaster Recovery Committee should ensure that the document is audited periodically (say once every quarter) against the present components in the organization. Another way to achieve this is to ensure that the committee is notified of any change that happens to any system/component in the organization so that the committee may update the document accordingly.

【Key words】

rehabilitation n. 修复；复兴
encompass vt. 围绕，包围；包含或包括某事物；完成
liquefaction n. 液化（作用）
floodway n. 分洪河道，泄洪渠，泄洪道
basin n. 盆；盆地；流域
vulnerability n. 弱点，攻击；易伤性
incorporate vi. 包含；吸收；合并
building code 建筑规范
flood-prone 易受水淹的
reconstitution mechanism 调整机制
mock drill 演习，演练
prioritize vt. 按重要性排列，划分优先顺序；优先处理
entity n. 实体；实际存在物
criticality n. 危险程度；临界
relocation n. 迁移；再定位
aftereffect n. 后果；事后影响
documentation n. 记录；参考资料；文献资料
periodic adj. 周期的；定期的
audit vt. 审计，查账；旁听

【Translation】

Disaster mitigation and rehabilitation have become some of the most pressing topics for discussion in geotechnical engineering and related professions for many years. Disaster management encompasses diverse topics such as natural disasters (tsunamis, earthquakes, landslides, etc.), mitigation (early warning and prediction systems, hazard mapping, risk analysis, coastal protection works, etc.), rehabilitation and reconstruction (difficult soils and ground treatment, design against earthquakes and other natural disasters, etc.), and many others, including soil dynamics, liquefaction, stability, and environmental protection.

防灾减灾和灾后恢复工作已经成为在岩土工程和相关专业领域讨论了许多年的一些最紧迫的话题。灾害管理包括不同的主题，如自然灾害（海啸、地震、山体滑坡等）、灾情缓解（早期预警和预报系统、灾害地图绘制、风险分析、海岸防护工程等）、灾后恢复和重建（困难的土壤和地基处理、对抗地震和其他自然灾害的设计等），以及许多其他的方面，包括土壤动力学、土壤液化、稳定性和环境保护。

Disaster mitigation   防灾减灾

Disaster mitigation measures are those that eliminate or reduce the impacts and risks of hazards through proactive measures taken before an emergency or disaster occurs.

防灾减灾措施是指在紧急情况或灾难发生前，通过采取主动措施消除或减少灾害的影响和风险的那些措施。

One of the best known examples of investment in disaster mitigation is the Red River Floodway. The building of the Floodway was a joint provincial/federal undertaking to protect the City of Winnipeg and reduce the impact of flooding in the Red River Basin. It cost \$60 million to build in the 1960s. Since then, the floodway has been used over 20 times. Its use during the 1997 Red River Flood alone saved an estimated \$6 billion. The Floodway was expanded in 2006 as a joint provincial/federal initiative.

在防灾减灾投资中最著名的例子之一是红河泄洪渠。修建红河泄洪渠是为了保护温尼伯市和减少洪水在红河盆地影响的一项省/联邦共建的工程。20世纪60年代时建造耗资6000万美元。从那时起，该泄洪渠已经被使用超过了20次。仅仅在1997年的红河洪水那一次应用中就节省了约60亿美元。该泄洪渠作为省/联邦共建项目在2006年进行了扩建。

All-hazards approach   危险全覆盖方法

An all-hazards emergency management approach looks at all potential risks and impacts, natural and human-induced (intentional and non-intentional) to ensure that decisions made to mitigate against one type of risk do not increase our vulnerability to other risks.

危险全覆盖应急管理方法着眼于所有潜在的风险和影响，自然的和人为诱导的（有意的和无意的），以确保为减轻一种风险所做的决定不会增加我们在面对其他风险时的脆弱性。

Types of disaster mitigation 防灾减灾类型

Disaster mitigation measures may be structural (e. g. flood dikes) or nonstructural (e. g. land use zoning). Mitigation activities should incorporate the measurement and assessment of the evolving risk environment. Activities may include the creation of comprehensive, pro-active tools that help decide where to focus funding and efforts in risk reduction.

防灾减灾措施可能是工程结构型的（比如防洪堤）或者非工程结构型的（比如土地用途分区制）。防灾减灾行动应该与不断变化的风险环境的监测和评估相结合。防灾减灾行动应该包括创建全面和积极的工具，用来帮助决定在减少风险方面该集中资助和努力的方向。

Other examples of mitigation measures include:

其他防灾减灾措施的例子包括：

- Hazard mapping
  危险地图绘制
- Adoption and enforcement of land use and zoning practices
  土地用途和分区使用的采用和执行
- Implementing and enforcing building codes
  实施和执行建筑规范
- Flood plain mapping
  洪水漫滩图绘制
- Reinforcedcity safe rooms
  加固城市安全用房
- Burying of electrical cables to prevent ice build-up
  掩埋电线以防止其结冰
- Raising of homes in flood-prone areas
  抬高在易受洪水影响地区房屋的地势
- Disaster mitigation public awareness programs
  防灾减灾公众意识计划
- Insurance programs
  保险计划

Recovery from disasters 灾后恢复

Disasters are inevitable but mostly unpredictable, and they vary in type and magnitude. The best strategy is to have some kind of disaster recovery plan in place, to return to normal after the disaster has struck.

Chapter 7　Disasters in Geotechnical Engineering　岩土工程灾害

灾害是不可避免的，且大多是不可预测的，灾害类型多变、大小不同。最好的策略是时刻准备好某种灾后恢复计划，在灾难发生后能够尽快恢复正常。

In the event of a large-scale natural disaster **where** response and recovery costs **exceed what** individual provinces and territories could reasonably **be expected to** bear on their own, the central government has to provide necessary assistances **including** financial support to the provincial and territorial governments at the first time. Assistance is paid to the province or territory—not directly to individuals or communities. The provincial or territorial governments design, develop and deliver disaster financial assistance, determining the amounts and types of assistance that will be provided to those who have experienced losses.

在发生大规模的自然灾害的事件时，应急响应和恢复的费用超过个别省份和地区能合理地自己承担的费用，中央政府必须提供必要的援助，包括在第一时间向省级和地区政府提供财政支持。财经援助款是拨给省部或地区的，而不是直接拨给个人或社区。省级或地区政府设计、开发和提供灾害性财政援助，确定援助的数量和种类，并提供给那些遭受了损失的人。

Every business disaster has one or more causes and effects. The causes **can be natural or human or mechanical** in origin, **ranging from** events such as a tiny hardware or software component's malfunctioning **to universally recognized** events such as earthquakes, fire and flood. Effects of disasters range from small interruptions to total business shutdown for days or months, even fatal damage to the business.

每一次灾难都有一种或多种原因和影响。这些原因可能源自自然的或人为的或机械的，可以是一个小的硬件或软件部件的失灵的小事件到广为人知的如地震、火灾和洪水等大事件。灾害的影响范围可从小型中断到持续数天或数月的全面关停，甚至对企业造成致命损害。

The process of preparing a disaster recovery plan begins by identifying these causes and effects, analyzing their likelihood and severity, and ranking them in terms of their business priority. The ultimate results are a formal assessment of risk, a disaster recovery plan that includes all available recovery mechanisms, and a formalized Disaster Recovery Committee that has responsibility for rehearsing, carrying out, and improving the disaster recovery plan.

准备灾后恢复计划的工作过程是，首先要确认灾害的原因和影响、分析其可能性和严重性，并根据其生产上的优先级进行排序。最终的结果包含一份正式的风险评估，一个包括所有可行的恢复机制的灾后恢复计划，还有一个负责排练、实施并改善灾后恢复计划的正规化的灾后恢复委员会。

The disaster recovery plan should: ①identify and classify the threats/risks that may lead to disasters; ②define the resources and processes that ensure business continuity during the disaster; ③define the reconstitution mechanism to get the business back to normal from the disaster recovery state, after the effects of the disaster are mitigated. An effective disaster recovery plan plays its role in all stages of the operations as depicted above, and it is continuously improved by disaster recovery mock drills and feedback capture processes.

灾后恢复计划应该：①对灾害可能导致的威胁/风险进行确定和分类；②明确灾害期间可保证企业继续经营的资源和办法；③在灾害的影响减缓后，明确能将企业生产从灾后恢复状态调整回正常状态的重建机制。一个有效的灾难恢复计划会在如上所述的各个实施阶段发挥作用，它能通过灾难恢复模拟演习和反馈意见收集流程不断地获得改进。

The methods and procedures involved in the disaster recovery planning process 灾后恢复计划过程中所涉及的方法和步骤

(1) The first step in planning recovery from unexpected disasters is to identify the threats or risks that can bring about disasters by doing risk analysis.

(1) 计划从突发灾害中恢复的第一步是要通过风险分析找出灾害可带来的威胁或风险。

(2) When evaluating risks, it is recommended to categorize them into different classes to accurately prioritize them. In general, risks can be classified in the following five categories: External Risks, Facility Risks, Data Systems Risks, Departmental Risks, Desk-Level Risks.

(2) 在评估风险时，建议将它们分为不同等级以便准确区分其优先顺序。在一般情况下，风险可以分为以下 5 类：外部风险、设备风险、数据系统风险、部门风险、办公桌层面风险。

(3) Once the evaluation of the major risk categories is completed, it is time to score and sort all of them, category by category, in terms of their likelihood and impact. The scoring process can be approached by preparing a score sheet.

(3) 一旦完成对主要风险类别的评估，就按其可能性和影响对它们进行评分和分类的时候了。评分过程可以通过准备一份评分表来实现。

(4) Once the disaster risks have been assessed and the decision has been made to cover the most critical risks, the next step is to determine and list the likely

effects of each of the disasters. These specific effects are what will need to be covered by the disaster recovery process.

（4）一旦已经评估了灾害风险，并已经做出承担最大风险的决定，下一步便是确定和列出每一种灾害可能带来的所有影响。这些具体的影响便是需要由灾后恢复过程来承担的内容。

(5) Once the list of affected entities is prepared and each entity's business criticality and failure tendency is assessed, it is time to analyze various recovery methods available for each entity and determine the best suitable recovery method for each. This step defines the resources employed in recovery and the process of recovery.

（5）一旦准备好了受影响对象的清单，并对每一对象的业务关键性和破坏倾向进行了评估，就要分析各种可用于每项影响事件的恢复方法，并为每项影响事件确定最合适的恢复方法。这一步决定了在恢复过程中所采用的资源和恢复的过程。

(6) Disaster recovery operations and procedures should be governed by a central committee. This committee should have representation from all the different company agencies with a role in the disaster recovery process, The Disaster Recovery Committee creates the disaster recovery plan and maintains it. During a disaster, this committee ensures that there is proper coordination between different agencies and that the recovery processes are executed successfully and in proper sequence.

（6）灾后恢复的行动和程序应由某个中心委员会来管理。该委员会应包括来自公司所有不同部门的代表，并且在灾后恢复过程中承担各自的作用，灾后恢复委员会要制定灾后恢复计划并落实好它。灾难期间，该委员会要确保不同机构之间良好协调，灾后恢复行动能按照正确的顺序成功实施。

The different phases of disaster recovery　　灾后恢复的不同阶段

Disaster recovery happens in the following sequential phases.

灾后恢复根据以下次序分阶段进行。

(1) Activation Phase. In this phase, the disaster effects are assessed and announced. The activation phase involves:

（1）启动阶段。在这个阶段，要评估并宣布灾害的影响。启动阶段包括：

- Notification procedures

- Damage assessment
  灾情估计
- Disaster recovery activation planning
  灾后恢复启动计划

(2) Execution Phase. In this phase, the actual procedures to recover each of the disaster affected entities are executed. Business operations are restored on the recovery system. Recovery operations start just after the disaster recovery plan has been activated, appropriate operations staff have been notified, and appropriate teams have been mobilized. The activities of this phase focus on bringing up the disaster recovery system. Depending on the recovery strategies defined in the plan, these functions could include temporary manual processing, recovery and operation on an alternate system, or relocation and recovery at an alternate site.

(2) 执行阶段。在这一阶段，执行用于恢复每一项受灾害影响事件的实际程序。企业经营活动在恢复系统中获得恢复。在灾害恢复计划已经启动后各种恢复行动业已开始，相应的工作人员已经通知到位，而且相应的团队也已经调动起来。这一阶段的行动重点是放在建立灾害恢复系统上。根据计划中确定的恢复策略，可以包括在一个可替代的系统中进行临时人工处理、恢复和操作，或在一个备用的地点进行安置和恢复等功能。

(3) Reconstitution Phase. In this phase the original system is restored and execution phase procedures are stopped. In the reconstitution phase, operations are transferred back to the original facility once it is free from the disaster aftereffects, and execution-phase activities are subsequently shut down. If the original system or facility is unrecoverable, this phase also involves rebuilding. Hence the reconstitution phase may last for a few days to few weeks or even months, depending on the severity of destruction and the site's fitness for restoration. As soon as the facility, whether repaired or replaced, is able to support its normal operations, the services may be moved back. The execution team should continue to be engaged until the restoration and testing are complete.

(3) 重建阶段。在这个阶段，原来的系统得到恢复，而执行阶段的程序也被终止。在重建阶段，一旦摆脱灾害的事后影响，经营活动被转移到原有的设施中，而执行阶段的行动也逐步终止。如果原有的系统或设备是不可恢复的，这一阶段还应包括重建。因此，重建阶段可能会持续数天、数周甚至数月，而这取决于灾害破坏的严重程度和场地对恢复的适宜性。不管是被维修好了还是进行了更换，只要场地设施能够支持正常的运行，各个服务部门就可以回去办公。执行团队应继续工作直至恢复作业和测试工作完成为止。

What information the disaster recovery plan should contain
灾后恢复计划应包含哪些信息

The outcome of the disaster recovery planning process is the disaster recovery plan document. During an emergency, this document will be the primary source of information for disaster recovery procedures.

灾后恢复计划过程的结果是灾后恢复计划文件。紧急情况期间，这份文件将是开展灾后恢复程序所需信息的主要来源。

The disaster recovery plan document is the only reliable source of information for the disaster recovery during an emergency. It should be very easily readable, with simple and detailed instructions. Some of the contents that need to be in this document are as follows:

灾后恢复计划文件是发生紧急情况期间开展灾后恢复所需信息的唯一可靠来源。这份文件应该是通俗易懂的，给出简单和详细的说明。以下是该文件中需要的一些内容：

Document Information/Purpose/Scope/Assumptions/Exclusions/System Description/Roles and Responsibilities/Contact Details/Activation Procedures/Execution Procedures/Reconstitution Procedures.

文件信息/使用目的/适用范围/基本假设/除外条款/系统描述/角色和职责/联系方式/启动流程/执行流程/重建流程。

How to maintain the disaster recovery plan 如何维护灾后恢复计划

The disaster recovery plan document needs to be kept up to date with the current organization environment. A plan that is not updated and tested is as bad as not having a plan at all because during emergencies, the document may be misleading. The following are recommended for maintenance of the plan documentation.

灾后恢复计划文件需要与公司当前的环境保持一致。一项计划如果不及时更新和测试的话，跟完全没有计划是一样糟糕的，因为在紧急情况下，这一文件还可能出现误导。为保持计划文件更新特提出以下建议。

- Periodic mock drills: The disaster recovery plan should be tested from time to time using scheduled mock drills. A drill usually will not affect active operations; however, if it is known that operations will be affected, the drill should be carefully scheduled such that the effect is minimal and is done during a permissible window. These activities should be regarded similarly to regular equipment maintenance activities that require operations downtime. The experience of the mock drill should be updated into the disaster recovery plan document.

定期模拟演练：使用定期模拟演练，时不时对灾后恢复计划进行测试。一次演练通常不会影响实际运行工作，但是，如果知道实际运行会受到影响，就要仔细安排好演练，使其影响最小化，或在允许的窗口期来进行。这些活动应被视为类似于对正常的设备进行维护保养工作，需要一定的运行停机时间。来自模拟演练的经验应及时更新到灾后恢复计划文件中。

- Experience capture: The best testing the document will undergo is when an actual disaster happens, and the lessons learned during the disaster recovery are valuable for improving the plan. Hence the Disaster Recovery Committee should ensure that the experience gets captured as lessons learned and the document gets updated accordingly.

获取经验：一份文档接受的最佳检验是在一个真正的灾难发生之时，在灾后恢复期获得的经验教训对改善计划来说是十分宝贵的。因此，灾后恢复委员会应确保能够获得必要的经验，同时也要吸取教训并使方案得到相应的更新。

- Periodic updates: Technologies, systems, and facilities that the plan covers may change over time. It is important that the disaster recovery plan document reflect the current information about the components it covers. For this purpose, the Disaster Recovery Committee should ensure that the document is audited periodically (say once every quarter) against the present components in the organization. Another way to achieve this is to ensure that the committee is notified of any change that happens to any system/component in the organization so that the committee may update the document accordingly.

定期更新：灾后恢复计划中所涵盖的技术、系统和设施可能随着时间发生改变。因次，灾后恢复计划方案反映它所包含的各个方面的最新信息是非常重要的。为了这个目的，灾后恢复委员会应确保计划文档会针对公司现行各个部门的情况进行定期（比如说季度性的）审核。达到这一目标的另一种方法，则是确保委员会能获知在机构中任何系统/部门发生的改变，委员会进而可以相应地更新该文档。

【Important sentences】

1. In the event of a large-scale natural disaster where response and recovery costs exceed what individual provinces and territories could reasonably be expected to

bear on their own, the central government has to provide necessary assistances including financial support to the provincial and territorial governments at the first time.

where 引导地点状语从句; what 引导宾语从句; be expected to do sth. 被期待去做某事; including 现在分词，补充说明包括的内容。

2. The causes can be natural or human or mechanical in origin, ranging from events such as a tiny hardware or software component's malfunctioning to universally recognized events such as earthquakes, fire, and flood.

can be natural or human or mechanical 可能是自然的或人为的或机械的; ranging from…to… 从……到……，表范围; universally recognized 过去分词作定语修饰 events。

Chapter 8
Numerical Technologies and Geotechnical Engineering

数值技术与岩土工程

8.1 Basic concepts in numerical technologies
数值技术基本概念

【Text】

A number of numerical methods of analysis have been developed over the past few decades. In Recent years numerical methods have continued to expand and diversify into all the major fields of scientific and engineering studies. They have become popular due to rapid advancements in computer technology and its availability to engineers. They provide a viable alternative to physical models that can be expensive and time consuming.

Before the advent of computers, the rock structures were designed largely based on rules of thumb, experience and a trial and error procedure. Rules of thumb are invariably based on the past experience of the designer. They usually tend to be over safe and are basically applicable to the situations similar to the ones for which they were developed.

Engineers of today are, many a time faced with problems for which no past experience is available. It's also difficult to "teach" past experience. The civil and mining engineering construction is usually a "one off" situation every time. The increased consciousness amongst the public regarding safety and economy has led the engineers to seek more rational solutions to the problems in rock mechanics related to civil and mining engineering. Analytical or "closed form" solutions are available for simpler situations or can be developed. However, they can in most cases be developed assuming rock as a linear elastic material which is a very drastic simplification. Numerical methods have, therefore, become very popular for solving problems in rock mechanics.

A number of numerical methods are available for solving problems of load deformation. By the term load-deformation, we mean a problem in which a rock mass of arbitrary shape (this includes openings of arbitrary shape) is subjected to loads due to self weight, external forces, in situ stresses, temperature changes, fluid pressure, pre-stressing, dynamic forces, etc. and we seek to find the deformation, strains and stresses throughout the rock mass.

The solution of this problem must satisfy the following:

(a) Equilibrium
(b) Strain compatibility

(c) Stress-strain relations of the rock mass
(d) Boundary considerations of tractions (forces) and deformations (conditions of fixity)

All numerical methods satisfy the conditions (a), (b) and (d) in almost a routine manner. Stress-strain relations for rock masses (c) is a wide subject in itself and perhaps most crucial on which the usefulness of the solution depends.

【Key words】

time consuming　耗费时间的，旷日持久的
rules of thumb　经验法则
trial and error　反复试验；试差法，试错法
oversafe *adj.* 过于安全的
arbitrary *adj.* 随意的，任性的
equilibrium *n.* 平衡，均势
traction *n.* 拖拉，牵引力，附着摩擦力

【Translation】

A number of numerical methods of analysis have been developed over the past few decades. In Recent years numerical methods have continued to expand and diversify into all the major fields of scientific and engineering studies. **They have become popular due to rapid advancements in computer technology and its availability to engineers.** They provide a viable alternative to physical models that can be expensive and time consuming.

在过去的几十年里，我们已经开发了许多数值分析方法。近年来，数值算法持续拓展和多样化，深入科学和工程研究的所有重要领域。由于计算机技术的迅速发展，加上工程师们能方便获取，数值算法已经变得很普及。它们能够为既贵又费时的物理模型提供一条可行的替代途径。

Before the advent of computers, the rock structures were designed largely based on rules of thumb, experience and a trial and error procedure. Rules of thumb are invariably based on the past experience of the designer. **They usually tend to be over safe and are basically applicable to the situations similar to the ones for which they were developed.**

在计算机出现之前，岩体结构主要根据单纯的经验法、借鉴以往经验以及反复试算的方法进行设计。单纯的经验法死板地依赖设计人员过去的经验。它们通常会过于安全，且基本上只适用于和他们所完成的项目具有相似条件的情况。

8.1 Basic concepts in numerical technologies 数值技术基本概念

Engineers of today are, many a time faced with problems for which no past experience is available. It's also difficult to "teach" past experience. The civil and mining engineering construction is usually a "one off" situation every time. **The increased consciousness amongst the public regarding safety and economy has led the engineers to seek more rational solutions to the problems in rock mechanics related to civil and mining engineering.** Analytical or "closed form" solutions are available for simpler situations or can be developed. However, they can in most cases be developed **assuming rock as a linear elastic material** which is a very drastic simplification. Numerical methods have, therefore, become very popular for solving problems in rock mechanics.

今天的工程师，很多时候所面临的问题并没有现成的经验可取。用"教"的办法去传授以往的经验也很困难。土木与采矿工程施工每一次通常都是"一次性"的工程。公众对于安全和经济意识的提高，促使工程师们为土木和采矿工程相关的岩石力学问题，寻求更合理的解决方案。分析式或"封闭式"的解决方案可用于较为简单的情况或可以对它们进行演绎。然而，大多数情况下，它们是在把岩石假定为线弹性材料条件下进行的演绎，而这一假定是很大程度上的简化。因此，数值算法在岩石力学问题的求解中已经变得非常普遍。

A number of numerical methods are available for solving problems of load deformation. By the term load-deformation, we mean a problem in which a rock mass of arbitrary shape (this includes openings of arbitrary shape) is subjected to loads due to self weight, external forces, in situ stresses, temperature changes, fluid pressure, pre-stressing, dynamic forces, etc. and we seek to find the deformation, strains and stresses throughout the rock mass.

有许多数值算法可用于解决加载-变形的问题。当我们用到加载-变形这一术语时，意指这一类问题：任意形状的岩体（包括任意形状的硐室）会受到由于自重、外部力量、原岩应力、温度变化、流体压力、预应力、动力等荷载作用，我们要找出整个岩体中的变形、应变和应力。

The solution of this problem must satisfy the following:

这个问题的解答必须满足以下条件：

(a) Equilibrium
力的平衡
(b) Strain compatibility
应变相容性
(c) Stress-strain relations of the rock mass

岩体的应力-应变关系
(d) Boundary considerations of tractions (forces) and deformations (conditions of fixity)
满足边界上牵引力（力）和变形（固定条件）的边界条件

All numerical methods satisfy the conditions (a), (b) and (d) in almost a routine manner. Stress-strain relations for rock masses (c) is a wide subject in itself and perhaps most crucial on which the usefulness of the solution depends.

所有的数值算法差不多自然而然地能满足条件（a）、（b）和（d）。而岩体的应力-应变关系，即条件（c）本身是一个涉及面很广且很重要的问题，而答案的有效性则取决于这一关系。

【Important sentences】
1. They have become popular due to rapid advancements in computer technology and its availability to engineers.
   due to 因为；rapid advancements 迅速的发展，rapid 作 advancements 的定语；in computer technology 作状语；and 表并列。
2. They usually tend to be over safe and are basically applicable to the situations similar to the ones for which they were developed.
   tend to，固定句型，"趋向于"；over safe 过于安全；basically 作 applicable 的状语；for which they were developed 作 ones 的定语从句。
3. The increased consciousness amongst the public regarding safety and economy has led the engineers to seek more rational solutions to the problems in rock mechanics related to civil and mining engineering.
   increased consciousness amongst… 在……方面意识的提高；seek more rational solutions 寻求更合理的解决方案；related to sth. 与 sth. 相关的。
4. assuming rock as a linear elastic material
   assuming A as B 把 A 假定为 B。

8.2 Numerical methods and their advantages
常见数值方法及其优越性比较

【Text】
There are mainly three numerical methods which have been used in the problems of rock mechanics.

- FEM：The Finite Element Method

8.2 Numerical methods and their advantages 常见数值方法及其优越性比较

- BEM: The Boundary Element Method
- DEM: The Discrete Element Method

FEM

This is the most popular method in engineering sciences. It has been applied to a large number of problems in widely different fields.

A large part of the finite element program can remain as a "black box" to the user and even a beginner can obtain interesting results with minimal effort. It does not mean that the method is easy and no experience is required in solving engineering problems. On the contrary, to make use of the full potential of the method and interpret the results of the calculation, considerable expertise is required.

The method essentially involves dividing the body in smaller "elements" of various shapes (triangles or rectangles in two-dimensional cases and tetrahedrons or "bricks" in three-dimensional cases) held together at the "nodes" which are corners of elements.

The more the number of elements used to model the problem, the better approximation to the solution is obtained. Displacements at the nodes are treated as unknowns and are calculated. Stress is calculated at one or more points inside each of the elements. Each element can have different material properties.

The major disadvantage of the method is that considerable effort is required in preparing data for the problem. This is particularly crucial in three-dimensional problems. The method is also expensive in computer time.

A large set of simultaneous equations have to be solved to obtain solutions. The computer time goes up further if the problem is nonlinear. For a nonlinear problem, the sets of simultaneous equations are required to be solved a number of times.

In spite of the above disadvantages, FEM has been extremely popular with geotechnical engineers. Its strength lies in its generality and flexibility to handle all types of loads, sequences of construction, installation of supports, etc.

BEM

This method is becoming increasingly popular. It lacks the generality and flexibility of the FEM. It is not so easily understandable and requires a higher level of understanding of mathematical complexities.

In this method only the surface of the rock mass to be analyzed needs to be discretized, i. e. divided into smaller patches. Thus for two-dimensional situations line elements at the boundary represent the problem, while for fully three-dimensional problems, surface elements are required. The data preparation is relatively simple. Whenever there is a change of material properties, the surface defining the separation has to be discredited. Thus, if there are a number of layers of different materials, data preparation can still become complex.

BEM appears to be a very efficient method for homogenous, linear elastic problems, particularly in three dimensions. For complex nonlinear material laws, advantages of the method are considerably diminished.

The method makes use of certain closed form relations of what may be called "elementary" problems. These solutions frequently contain trigonometric and logarithmic terms which slow down the computations.

Recognizing the advantages and disadvantages of the two methods, viz. FEM and BEM, many researchers have combined the two methods. This is coupled FEM/BEM method in which for a certain region (usually close to an opening or some other feature of interest) FE discretization is used, while for other regions BE discretization is adopted.

DEM

This method is based on treating the rock mass as a discontinuum rather than continuum, as in the case of Finite Element and Boundary Element Methods. When loads applied, the changes in contact forces are traced with time.

In the earlier versions of the method rigid spherical balls or discs were used as elements. The equation of dynamic equilibrium for each element is repeatedly solved till the laws of contacts and boundary conditions are satisfied.

In the recent versions of the method, the elements can be of arbitrary shape as in the Finite Element Method. They can also be deformable. Complex constitutive laws can be used. The element can be split up based on the assumed fracture criterion during the calculation process without any external intervention.

Thus, the method is extremely powerful. There are, however, several drawbacks.

(1) The parameters required for the description of material behaviour are required to be chosen quite carefully in addition to certain additional parameters like

the damping of the system.

(2) Computation time required to solve even simple problems can be excessive.

At present, the method appears to be extremely useful in explaining the deformation and failure of rock masses qualitatively and provides a valuable insight into the failure mechanism. More experience is, however, required for it to be an acceptable tool of analysis in practice.

【Key words】
tetrahedron *n*. ［晶体］四面体
node *n*. 节点
simultaneous equation 联立方程
discretize *vt*. 使离散；离散化
line element 线元
surface element 面积元
homogenous *adj*. ［生物］同质的；同类的
trigonometric *adj*. ［数］三角法的；三角学的
logarithmic *adj*. 对数的
discontinuum *n*. 不连续体
rigid spherical ball 刚性球体
fracture criterion 断裂准则
qualitatively *adv*. 定性地；从品质上讲，质量上

【Translation】
There are mainly three numerical methods which have been used in the problems of rock mechanics.

用来解决岩石力学问题的数值算法主要有3种。

- FEM：The Finite Element Method
  有限元方法
- BEM：The Boundary Element Method
  边界元方法
- DEM：The Discrete Element Method
  离散元方法

FEM 有限元方法

This is the most popular method in engineering sciences. It has been applied to a large number of problems in widely different fields.

有限元方法是在工程科学中最通用的方法，已经广泛地用于不同领域大量问题的分析求解。

A large part of the finite element program can remain as a "black box" to the user and even a beginner can obtain interesting results with minimal effort. It does not mean that the method is easy and no experience is required in solving engineering problems. **On the contrary, to make use of the full potential of the method and interpret the results of the calculation, considerable expertise is required.**

对用户而言，有限元程序中的很大一部分可以作为"黑箱子"存在，甚至对于一个初学者而言，只要小的付出就能得到有趣的结果。但这并不意味着该方法是简单容易的，在用它来解决工程问题时不需要经验。相反，为了充分利用该方法的潜能、解读计算结果，需要相当程度的专门知识。

The method essentially involves dividing the body in smaller "elements" of various shapes (triangles or rectangles in two-dimensional cases and tetrahedrons or "bricks" in three-dimensional cases) held together at the "nodes" which are corners of elements.

该方法实质上是将物体分解成各种形状的较小的"单元"（如二维条件下的三角形或矩形，或三维条件下的四面体或"块体"），各单元在"节点"处联结在一起，"节点"通常是各单元的棱角。

The more the number of elements used to model the problem, **the better** approximation to the solution is obtained. Displacements at the nodes **are treated as unknowns** and are calculated. Stress is calculated at one or more points inside each of the elements. Each element can have different material properties.

模拟问题所用的单元数量越多，所获解答的近似程度越高。节点处的位移被视为待求未知量来进行计算求解。每个单元内的一个或多个点处的应力要进行计算。每个单元可以有不同的材料属性。

The major disadvantage of the method is that considerable effort is required in preparing data for the problem. This is particularly crucial in three-dimensional problems. The method is also expensive in computer time.

该方法的主要缺点是在为分析问题准备数据时需要耗费可观的精力。这一点在解决三维问题时尤其严重。另外，该方法的计算机用时量长。

A large set of simultaneous equations have to be solved to obtain solutions. The computer time **goes up further** if the problem is nonlinear. For a nonlinear problem, the sets of simultaneous equations are required to be solved a number of times.

为了获得解答，需对大规模的联立方程组进行求解。如果问题是非线性的话，计算机用时会进一步上升。对于非线性问题，联立方程组需要进行多次求解。

In spite of the above disadvantages, FEM has been **extremely popular with** geotechnical engineers. Its strength lies in its generality and flexibility to handle all types of loads, sequences of construction, installation of supports, etc.

尽管存在上述缺点，有限元法对于岩土工程师而言已经极其通用。有限元法的优势在于它处理各种荷载、施工顺序、支护安装等问题时的通用性和灵活性。

BEM 边界元方法

This method is becoming increasingly popular. It lacks the generality and flexibility of the FEM. It is not so easily understandable and requires a higher level of understanding of mathematical complexities.

边界元方法正变得越来越流行。该方法缺少像有限元方法的通用性和灵活性。它也不是那么容易理解，需要更高的理解复杂数学问题的能力和水平。

In this method only the surface of the rock mass to be analyzed needs to be discretized, i.e. divided into smaller patches. Thus for two-dimensional situations line elements at the boundary represent the problem, while for fully three-dimensional problems, surface elements are required. The data preparation is relatively simple. Whenever there is a change of material properties, the surface defining the separation has to be discredited. Thus, if there are a number of layers of different materials, data preparation can still become complex.

在这种方法中，只需对所分析的岩体的表面进行离散化，即将其分解为较小的面单元。因此，对于二维情况而言，边界上的线元素对应待解决的问题，而对于完全的（真）三维问题，则需要用面单元。数据准备工作相对简单。每当材料属性有变化时，用于区分这一变化的分界面就必须进行离散化。因此，如果存在大量不同材料分层，数据准备工作也会变得复杂。

BEM appears to be a very efficient method for homogenous, linear elastic problems, particularly in three dimensions. For complex nonlinear material laws, advantages of the method are considerably diminished.

边界元方法对于各向同性的线弹性问题，特别是在三维的情况下，显然是一种非常有效的方法。面对材料复杂的非线性定律时，该方法的优点会明显受到制约。

The method makes use of certain closed form relations of what may be called "elementary" problems. These solutions frequently contain trigonometric and logarithmic terms which slow down the computations.

该方法会利用一些可被称为"基本"问题的固定式关系。这些解答通常包含三角函数和对数的内容，而这会减慢运算速度。

Recognizing the advantages and disadvantages of the two methods, viz. FEM and BEM, many researchers have combined the two methods. This is coupled FEM/BEM method in which for a certain region (usually close to an opening or some other feature of interest) FE discretization is used, while for other regions BE discretization is adopted.

许多研究者认识到了有限元和边界元两种方法的优缺点，已经将两者进行了有机结合。在有限元和边界元相结合的方法中，一些区域（通常是接近开口或其他一些特定区域）采用有限元进行离散化处理，而另外一些区域则采用边界元来进行离散化处理。

DEM 离散元方法

This method is based on treating the rock mass as a discontinuum rather than continuum, as in the case of Finite Element and Boundary Element Methods. When **loads applied**, the changes in **contact forces are traced with time**.

离散元方法以把岩体看成是非连续体而不是连续体为基础，这跟有限元和边界元方法中的情况是不一样的。当施加荷载时，接触力随时间而变化。

In the earlier versions of the method rigid spherical balls or discs were used as elements. The equation of dynamic equilibrium for each element is repeatedly solved till the laws of contacts and boundary conditions are satisfied.

在该方法的早期版本中，刚性球体或碟体被作为单元体。对各单元体的动态平衡方程进行反复求解，直到各单元体间的接触定律和边界条件获得满足。

In the recent versions of the method, the elements can be of arbitrary shape as in the Finite Element Method. They can also be deformable. Complex constitutive laws can be used. The element can be **split up based on the assumed fracture criterion during** the calculation process without any **external intervention**.

在该方法的近期版本中，与有限元方法一样，单元体可以是任意形状的。它们也是可变形的。能够使用复杂的本构关系。在计算过程中，单元可以在没有任何外部干预的情况下，根据假定的断裂准则进行裂解。

Thus, the method is extremely powerful. There are, however, several drawbacks.

因此，离散元方法是非常强大的。然而，它也有几点不足。

(1) The parameters required for the description of material behaviour are required to be chosen quite carefully in addition to certain additional parameters like the damping of the system.

(1) 除了某些如系统的阻尼系数的附加参数外，描述材料性质所需的参数需要进行相当谨慎的选择。

(2) Computation time required to solve even simple problems can be excessive.

(2) 即使要解决简单的问题，计算所需的时间也可能非常多。

At present, the method appears to be extremely useful in explaining the deformation and failure of rock masses qualitatively and provides a valuable insight into the failure mechanism. More experience is, however, required for it to be an acceptable tool of analysis in practice.

目前，该方法在定性解释岩体的变形和破坏方面显然是极其有效的，能为岩体破坏机理提供有价值的见解。但是，为了使它能够在实践中成为一种可接受的分析工具，还需要积累更多的经验。

【Important sentences】

1. On the contrary, to make use of the full potential of the method and interpret the results of the calculation, considerable expertise is required.
   On the contrary 相反；to make use of 为了要充分利用；considerable 相当大的；is required 表被动。
2. The more the number of elements used to model the problem, the better approximation to the solution is obtained.
   The more…the better…，固定句型，"越多越好"；approximation 接近，近似。
3. Displacements at the nodes are treated as unknowns and are calculated.
   be treated as 被视为；unknowns 作名词，待求未知量。
4. The computer time goes up further if the problem is nonlinear.
   goes up further 进一步上升；nonlinear 非线性的。
5. In spite of the above disadvantages, FEM has been extremely popular with

geotechnical engineers.

In spite of，固定短语，"尽管"；be extremely popular with 极其流行（适用）。

6. When loads applied, the changes in contact forces are traced with time.

loads applied 施加荷载；contact forces 接触力；be traced with time 随着时间的推移被追踪。

7. The element can be split up based on the assumed fracture criterion during the calculation process without any external intervention.

split up 分裂，分离；based on 基于，根据；assumed fracture criterion 假定的断裂准则；external intervention 外部干预，external 外部的，internal 内部的。

8.3 Future challenges 学科领域未来的挑战

【Text】

A number of powerful and versatile numerical methods are available to the engineer today.

The advent of fast and friendly personal computers has made it possible to solve complex problems which, in the past, could not be solved without mainframe machines and without considerable effort in terms of time and resources.

The speed of computation has increased a lot in the past decades. This trend is likely to continue. It is therefore apparent that numerical techniques will play an increasingly crucial role in the solution of engineering problems.

The main task of the engineer in applying the numerical methods to the problems of rock mechanics is in quantifying the mechanical response of rocks, rock joints, rock masses and rock support systems. This requires the background knowledge of models of standard material behavior such as elastic, elasto-plastic and elasto-viscoplastic.

The mathematical theories lead to "phenomenological laws" of material behavior, i.e. the relationships between stress and strain on an average basis.

Due to the advances in numerical methods and computing, there has been an exponential growth in the models of the behavior of engineering materials which are largely based on the theory of plasticity and viscoplasticity.

【Key words】

versatile *adj.* （指工具、机器等）多用途的；多才多艺的；多功能的

mainframe *n.* （大型电脑的）主机，中央处理机
quantify *vt.* 确定……的数量；量化
phenomenological law   现象学定律
exponential *adj.* 指数的

【Translation】

A number of powerful and versatile numerical methods are available to the engineer today.

如今，大量功能强大的和适用范围广的数值算法可供工程师选择。

The advent of fast and friendly personal computers **has made it possible to** solve complex problems **which，in the past，could not be solved without mainframe machines and without considerable effort** in terms of time and resources.

快速而方便的个人电脑的出现已经使得解决复杂的问题成为可能，而这在过去，要是没有主机，没有可观的时间和资源上的付出，是不可能解决的。

The speed of computation has increased a lot in the past decades. This trend is likely to continue. It is therefore apparent that numerical techniques will play an increasingly crucial role in the solution of engineering problems.

在过去的几十年中，计算速度已经获得提升，而这种趋势还很可能持续。因此，很明显，数值技术在解决工程问题上将起到越来越重要的作用。

The main task of the engineer in applying the numerical methods to the problems of rock mechanics is in **quantifying** the mechanical response of rocks，rock joints，rock masses and rock support systems. This requires the background knowledge of models of standard material behavior such as elastic，elasto-plastic and elasto-viscoplastic.

在运用数值方法来解决岩石力学问题方面，工程师的主要任务在于量化岩石、岩石节理、岩体和岩体支护系统的力学响应作用。而这要求了解诸如弹性、弹-塑性和弹-黏-塑性的标准材料性能模型的背景知识。

The mathematical theories lead to "phenomenological laws" of material behavior，i. e. the relationships between stress and strain on an average basis.

数学理论能导出材料性能的"现象学定律"，即以均值为基础的应力和应变关系。

Chapter 8 Numerical Technologies and Geotechnical Engineering 数值技术与岩土工程

Due to the advances in numerical methods and computing, there has been an exponential growth in the models of the behavior of engineering materials which are largely based on the theory of plasticity and viscoplasticity.

由于数值方法和计算能力的发展，工程材料特性模型出现了指数式增长，其中大多数是基于塑性和黏塑性理论的。

【Important sentences】

1. The advent of fast and friendly personal computers has made it possible to solve complex problems which, in the past, could not be solved without mainframe machines and without considerable effort in terms of time and resources.
   has made it possible to do sth. 使……成为可能；which 引导定语从句；in the past 做插入语；could not be solved without… and without… "要是没有……，没有……，是不可能解决的。"

2. in quantifying the mechanical response
   quantifying 量化。

Appended Pictures   附图

Articulated Trucks
铰接式卡车

Asphalt Pavers
沥青摊铺机

Cold Planers
冷铣刨机

Backhoe Loaders
反铲挖掘装载机

Compactors
压路机

Compact Track and Multi-terrain Loaders
紧凑轨道和多地形装载机

Dozers
推土机

Draglines
索斗控土机

Drills
钻孔

Electric Rope Shovels
电动绳铲

Excavators
挖掘机

Wheel Tractor Scrapers
轮式拖拉机式铲运机

Appended Pictures 附图 /227/

Forwarders
短材集运机

Hydraulic Mining Shovels
液压矿用挖掘铲

Highwall Miners
高边坡用采矿机

Knuckleboom Loaders
节状臂式装载机

Material Handlers
材料搬运机

Motor Graders
平地机

Appended Pictures 附图

Off-highway Trucks
非高速公路载重卡车

On-highway Trucks
高速公路载重卡车

Pipelayers
管道铺设车

Road Reclaimers
公路取料机

Skid Steer Loaders
滑移转向装载机

Telehandlers
伸缩臂叉车

Appended Pictures 附图 /229/

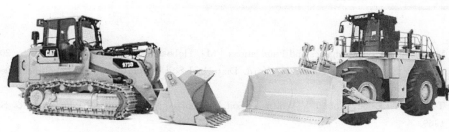

Track Loaders
履带式装载车

Wheel Dozers
轮式推土机

Wheel Excavators
轮式挖掘机

Wheel Loaders
轮式装载机

Underground Hard Rock Loaders
地下硬岩铲运车

References 参考文献

[1] Muni Budhu. Soil Mechanics and Foundations [M]. Hoboken: John Wiley & Sons, Inc., 2007.
[2] Andrew N Schofield, Thomas Telford. Disturbed Soil Properties and Geotechnical Design [M]. 2006.
[3] T William Lambe, Robert V Whitman. Soil Mechanics [M]. Hoboken: John Wiley & Sons, 1969.
[4] David Muir Wood. Soil Behavior and Critical State Soil Mechanics [C]. Cambridge: Cambridge University Press, 1990.
[5] Holtz R, Kovacs W. An Introduction to Geotechnical Engineering [M]. Upper Saddle River: Prentice-Hall Inc., 1981.
[6] Frederick S Merritt, M Kent Loftin, Jonathan T Ricketts. Standard Handbook for Civil Engineers [M]. 4th ed. New York: McGraw-Hill, 1995.
[7] R W Davidge. Mechanical Behavior of Ceramics. Cambridge Solid State Science Series, edited by D R Clarke et al., 1979.
[8] William D Callister. Fundamentals of Materials Science and Engineering [M]. Hoboken: John Wiley & Sons, 2004.
[9] Bird F, R Loftus. Loss Control Management [M]. Loganville, Ga.: Institute Press, 1976.
[10] Heinrich H W. Industrial Accident Prevention [M]. New York: McGraw-Hill, 1959.
[11] Hinze J. Construction Safety [M]. Upper Saddle River: Prentice-Hall Inc., 1997.
[12] Murphy D J. Safety and Health for Production Agriculture [R]. American Society of Agricultural Engineers, 1992.
[13] Roger L Brauer. Safety and health for engineers [M]. New York: Van Nostrand Reinhold, 1994.
[14] http://en.wikipedia.org/wiki/Geotechnical_engineering.